12/03

SECRETS OF THE ICE AGES

SECRETS
OF THE
ICE AGES

The role of the Mediterranean Sea in climate change

Robert G. Johnson
University of Minnesota

Glenjay Publishing
Minnetonka, Minnesota

Publisher's Cataloging-in-Publication
(Provided by Quality Books, Inc.)

Johnson, Robert G., 1922-
Secrets of the ice ages : the role of the Mediterranean Sea in climate change / Robert G. Johnson.
-- 1st ed.
p. cm.
Includes bibliographical references and index.
ISBN 0-9715630-0-4

1. Climatic changes. 2. Glacial epoch. 3. Geology--Mediterranean Sea. I. Title.

QC981.8.C5J64 2002 551.6
QBI02-200182

Library of Congress Control Number: 2002092043

Glenjay Publishing
12814 March Circle
Minnetonka, MN 55305
(952) 544 6979
glenjay@bitstream.net
SecretsoftheIceAges.com

Printed in the United States by:
Morris Publishing
3212 East Highway 30
Kearney, NE 68847

CONTENTS

The photograph on the front cover is a view of the tectonically-uplifted southeastern end of the First High Cliff, now eroding into the Caribbean Sea at Cave Bay on Barbados. This large coral-reef terrace was formed when sealevels were generally high between the last ice age, the Wisconsinan, and the maximum of the previous ice age, the Illinoian. The corals making up the reef grew during an interval of complex glacially-induced sealevel changes between 144,000 and 120,000 years ago. Photo by R.G. Johnson. The cover design was done by Wendy Skinner with assistance of Charles Hughes, Don Schultz, and Dave Stoken.

Preface

For the last 150 years, scientists have been searching for the cause of the cyclic appearance and disappearance of the great continental ice sheets of recent geologic time. This book is part of that search. It has been written for readers with a general interest in climate and Earth science as well as for scientists working in this field, and it is hoped that the conceptual model and ideas presented here will contribute significantly to our knowledge of the cause of ice-age climate change. And they should, because the new model is much more consistent with the physics of climate and the geological record than is the classical Milankovitch model. Customarily, the new concepts would be published as papers in archival journals, but these journals are not readily available to the ordinary reader. Furthermore, some of the climate-change events described in the book have not yet been exhaustively verified by conventional research. Ideas based on such events are seldom accepted by the journals' peer reviewers, even when supported by good evidence and sound inferences.

A generation ago exploratory Earth-science ideas had a better chance to be seen in print. But data trump ideas, and today, the pressures to publish data acquired in funded research leave little journal space for concepts that push the envelope of conventional wisdom. Investigation of such concepts is not favored in the allocation of available research funds, and the unconventional proposal is scorned by peer reviewers. However, new ideas are needed because the Milankovitch model of direct thermal control of glaciation is woefully inadequate, and, although much progress has been made, a half-century of intensive data acquisition has not yet yielded a satisfactory understanding of the most important features of Pleistocene climate change. The variations of Pleistocene climate and the variations of glacial-ice volume are inextricably coupled together, and much of the ice volume change occurs as a result of internal factors within the earth's climate system. However, climate proxies in the geological record do correlate with orbital insolation variations to some extent, and the correlations require an explanation.

Hence this book, which goes beyond the limitations of peer review to examine the role of the Mediterranean Sea in climate change. Among oceanographers and other Earth scientists the possibility of a significant Mediterranean influence on climate is a contentious matter, because it implies that a little of the conventional understanding of North Atlantic oceanic circulation is incorrect. Most peer reviewers favor only ideas and data that do not conflict with conventional facts. But facts are merely accepted ideas, and as Francis Crick of double helix fame has said: "Any theory that agrees with all the facts will be wrong because some of the facts will be wrong." Conversely, a theory may be quite correct but not agree with all the "facts."

The book is therefore an opportunity to present to a wide audience facts and arguments that support the importance of the Mediterranean Sea, and which might not otherwise appear in print. In doing so, it is necessary to examine past climates. However, there are no time machines to enable us to go back a hundred thousand years and observe the climates of the past, and we only "know" indirectly what has happened from clues in the geological records. Some important facts are consequently beyond the reach of observation and must be inferred if we are to make sense of climate history.

The central issue in Pleistocene climate change is the effect of long-term variations of incoming solar energy (insolation) on ice-age cycles. These variations occur because the tilt of Earth's polar axis is not constant and because orbital precession causes the summer season to occur when Earth is at varying distances from the sun. In the model advocated by Croll and Milankovitch, the ice-age cycles are supposed to be caused by the direct effect of received solar energy variations on climate warming and ice-sheet melting. This idea has been known loosely for some time as the Milankovitch hypothesis. Although many correlations between incident solar energy variations and ice-age changes are consistent with this hypothesis, the direct warming and melting effects cannot be the cause, and a satisfactory physical connection between the ice-age cycles and orbital variations has not been identified and accepted by the scientific community.

In this book, the Mediterranean Sea is the proposed link between variations of orbital parameter values and changes in the volume of glacial ice. In this very indirect and imperfect way, orbital variations have modulated the volume of glacial ice over the last three million years. The Mediterranean role and the associated new ideas discussed in the book are the result of disciplined analysis and are supported by considerable evidence. The wider acceptance of these ideas will not depend on any single new research result. In the earth sciences there is no "proof" of a disputed concept. Acceptance of any new idea depends, first of all, on thoughtful consideration of the idea itself, followed by acquisition of supporting data, if possible. Eventually the idea may fit into a broad web of related facts, and the skeptics - or most of them - become convinced.

The book is like a legal brief in which the evidence is outlined and the inferences are made to make the case for Mediterranean influence on ice-age cycles. It is not a review of climate-change research, however, and no attempt is made to mention all the significant contributions made by the many workers in this active field. In another decade or two the science will have advanced and some of the theoretical ideas discussed here may then have been validated - or replaced, for even the best hypotheses are but good approximations.

I have said very little about the role of modern global warming. That topic is covered well by others. However, warming does introduce an element of uncertainty into a timely ice-age prediction. But the arguments for the past and the predictions for the future are laid out here for all to see, and that great analog super-computer that we call the natural world will provide correct answers in due time.

Robert Glenn Johnson

Acknowledgments

There are many fine people who have made my activity in the field of earth science a very rewarding experience. Among them I am indebted to J.T. Andrews who a quarter of a century ago encouraged my earliest attempts to understand ice-age climate mechanisms. B.T. McClure shared an early interest in climate change and has done a careful reading of the book manuscript. R.K. Matthews briefed me on his experience in Barbados, and S.-E. Lauritzen and I explored the fascinating possibilities of the great Hudson Strait ice dams. R.L. Edwards educated me on coral basics. C. Paola cleared up puzzling questions on oceanic circulation. My colleague, Paul Weiblen, directed my attention to the "bi-polar seesaw" in northern climate, and our key insight into the switch to warm interglacial circulation was enabled by the efforts of J.-C Duplessy, who derived sea-surface temperatures and salinities for sediment core locations in the eastern North Atlantic. E. Ito's aid in preliminary efforts to organize this picture of ice-age climate variation is gratefully acknowledged, and R.L. Hooke's many glaciology seminars at his home and the fine hospitality of those occasions are unforgettable. A. Berger generously supplied his tabulated orbital parameters and insolation values, without which the neo-Milankovitch hypothesis of this book would not have been formulated. And I am particularly indebted to H.E. Wright, Jr., who over many years has supplied a stream of ideas and inspiration. Finally, the book would never have materialized without the long-time help of my wife, Elizabeth Gulliver Johnson, whose assistance has been invaluable in planning expeditions, obtaining maps and airline tickets, and operating the surveyor's rod when measuring elevations of terraces on Barbados.

RGJ

List of Illustrations

1

From Galileo to Milankovitch: A bit of history

Perhaps the most famous Renaissance scientist is Galileo Galilei, born in Pisa, Italy, in 1564. He was a man of many talents: an inventor, a musician, a mathematician, a physicist, and an astronomer. Above all, he was a man who entertained unconventional ideas based on new and sometimes unconventional facts. According to a legend, he dropped two heavy objects having different weights from the Leaning Tower of Pisa to successfully demonstrate that the speed of falling objects does not depend on their weights, contrary to the general opinion of the time. His most well known contribution to modern thought was his promotion of the idea of a heliocentric planetary system in which the Earth is only one of many planets orbiting around the sun.

Nicolaus Copernicus, an early Polish astronomer, first developed the heliocentric idea as a superior alternative to the Ptolemy concept of an Earth-centered universe. Galileo welcomed the idea, and made telescopic observations of the moons of Jupiter, which he likened to a miniature solar system, various phases of Venus, which he likened to phases of Earth's moon, and sunspots on what everyone had assumed to be a perfect and unblemished source of sunlight. He used these observations to support the Copernican theory of the solar system, and this put him at odds with the mainstream theology of the natural world in 1632. The theologians of the time viewed his discoveries as a threat to the Church, and his clash with the Inquisition in which he was forced to recant is an often-told story. Johannes Kepler, a German mathematician and a contemporary of Galileo, determined that the planetary orbits were elliptical rather than circular, and Kepler's laws of planetary motion accurately quantified the theory of Copernicus.

If Copernicus, Galileo and Kepler had not revolutionized our ideas on the relation between the sun and the earth, doubtless someone else would have done so. But that should not detract from their unique accomplishment in the realm of unconventional thought. The relevance of their discovery to changing ice-age climates lies in the varying shape of Earth's elliptical orbit around the sun, the variable tilt of Earth's polar axis, and the varying distance of Earth from the sun during summertime. These factors are discussed in Appendix A.

It is sufficient to point out here that these variations cause significant changes in the seasonal amount of received sunlight, or the insolation as it is usually called. These changes occur on timescales of tens and hundreds of thousands of years and are different at each latitude on the earth's surface. By variable heating of the land and the sea, in summer or winter, such changes can influence atmospheric and oceanic circulation, and thus can cause significant alterations in temperature, rainfall, and snowfall. But the earth's climate system, which consists of the oceans, the continents, the atmosphere, and the polar ice sheets, all interact in complex and chaotic ways, making it quite difficult to identify connections between the underlying solar driving force and the climate effects that we see in the geological records.

Only two centuries ago, the fact that climates have changed over long periods of time was hardly recognized. The knowledge of glacial geology and the historical perspective that we now have were lacking and the concept of an ice age was unknown. Today, it is quite clear to well-traveled and well-read northern people that large areas of North America and Europe were once covered by glacial ice. This picture of the world started to emerge early in the 19th century. One of the first scientists to publicize the fact that large land areas had been glaciated long ago was Louis Agassiz, a Swiss naturalist. Although skeptical at first, he became convinced by shrewd acquaintances and by his own observations made in the foothills of the Alps. There one often finds isolated and glacially-grooved rocks on land far below the small mountain glaciers of today. Clearly these glaciers had once carried the rocks through the foothills and out onto the plains. He and his associates observed similar striated rocks elsewhere in Europe and in the British Isles, and it became clear to them how widespread the glaciers had been in the past.

Agassiz became a professor at Harvard University in 1846 and was influential in igniting interest in the ice-age topic among scientists in the United States as well as in Europe. He and his American contemporaries soon recognized evidence of ice-age glaciation everywhere in the northern United States, from the loose boulder deposits on the bedrock in New England to the scattered boulders in the grass on the immense northern prairies west of the Mississippi River. Such rocks were always associated with the glacial till, the stones, clay, and sand that moving sheets of glacial ice dig out, mix up, transport, and deposit underneath the ice and at the ice-sheet edge. This giant ice sheet had once covered half of the North American continent (Fig 1). Agassiz and his fellow scientists also began to collect evidence of a vast inland lake that had formed temporarily on the plains of western Canada and northern Minnesota along the retreating edge of the ice sheet as it melted away. This lake at one time had an area more than three times as large as the present Lake Superior. In 1880, a few years after the death of Agassiz, Warren Upham, an eminent geologist, who had himself made major contributions in exploring the extent and history of the lake, named it Lake Agassiz in honor of the great contributions that Louis Agassiz made to the early knowledge of ice-age glaciation.

Eventually earth scientists recognized that at the last glacial maximum about 19,000 years ago, ice sheets with thicknesses as great as 3000 meters or more had covered the northern half of North America, all of Scandinavia, much of Northern Europe, and much of northern Siberia. This ice age was a long and cold event in much of the world. It started about 120,500 years ago when ice sheets began to grow in large areas of Baffin Island and northern Quebec in Canada. After tens of thousands of years of intermittent growth they extended far to the south over Europe and North America. Then, only about 18,000 years ago, the ice began a grand retreat to its Arctic origins, where only local glaciation at higher altitudes now remains. In the Northern Hemisphere during the present warm interglacial time, only the island of Greenland retains a large ice sheet. It is our inheritance from the last ice age, and we have it because it still snows a lot there. The low summer temperatures of the higher latitudes and larger amounts of precipitation due to the nearby warmer ocean allowed only a partial loss of the ice sheet during the last deglaciation.

The melting and retreat of these ice sheets left broad areas of land with a wide variety of terrains. On the plains of North America the passage of the ice tended to smooth

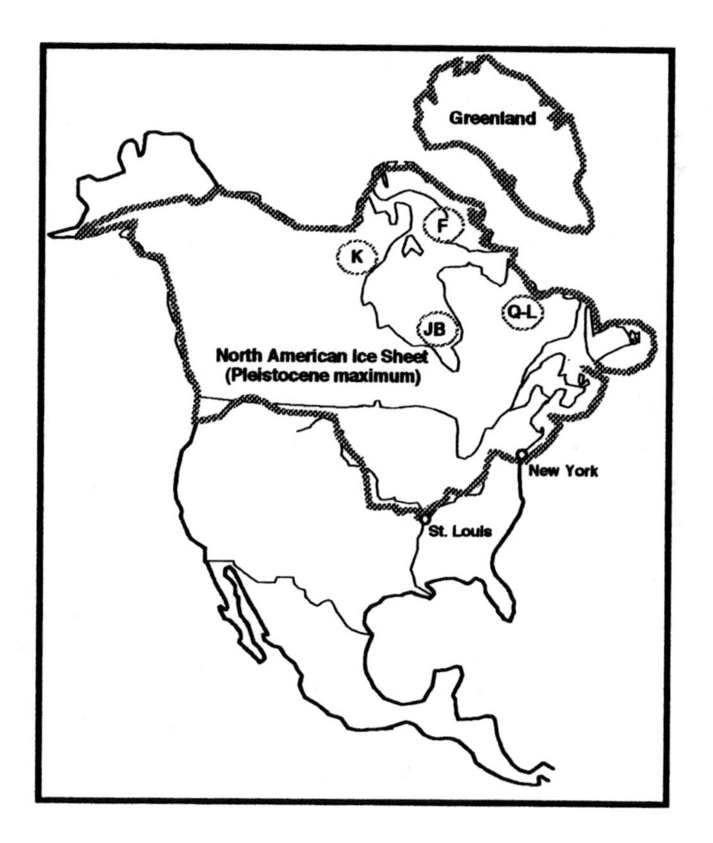

Figure 1: The North American ice sheet maximum during the last 2.5 million years. Glacial ice-domes: F - Foxe basin-Baffin, Q-L - Quebec-Labrador, JB - Hudson Bay-James Bay, K - Keewatin. At the most recent glacial maximum of the Wisconsinan ice age, the ice did not extend quite as far south as at the earlier maximum shown.

the landscape, leaving areas that became highly productive agricultural land in modern times. Where the front edge of the ice sheet was stationary, the land became covered with scenic hills of glacial till, the debris that had accumulated in the moving ice and was dumped at its front edge as the ice melted. Farther north the land was frequently scraped clean to the hard bedrock, and everywhere the landscape is now dotted with lakes both large and small, created by the combined effects of ice-sheet excavation and till deposition. In Europe and Scandinavia similar terrains occur, and in Siberia shallow lakes and swamps that were formed by the retreating ice are now filled with immense deposits of peat that have accumulated over the 10,000 years or more since the ice dominated the landscape.

As it became evident in the 19th century that great ice sheets had covered much of northern continents at least once, it became of considerable interest to find the cause for the widespread glaciation and its disappearance. One of the first to propose an astronomical cause for this was James Croll, who in 1875 suggested the hypothesis that the presence or absence of ice sheets depended directly on the melting rates of the glaciers, which changed with the orbital variation of summer sunlight (solar insolation). This variation is caused by the occurrence of Northern Hemisphere summer at different distances from the sun at points on the earth's elliptical orbit. In astronomical terms, this is the precession effect, which as defined, slowly shifts the position of the sun at the spring equinox westward among the background constellations of stars. The result is a movement of the position of summer along the earth's orbit, so that summer sometimes occurs close to the sun and at other times at greater distances. Croll's idea was thoroughly debated. His opponents pointed out that, although the ice age was worldwide, precession effects should cause ice ages to occur alternately about 11,500 years apart in the Northern and Southern Hemispheres. In Croll's day it was hard to answer this objection. Now we know that major ice-age climate oscillations occur simultaneously in both hemispheres because ice-age changes in North Atlantic oceanic circulation affect the oceans in the Southern Hemisphere also, and can cause a large expansion of the sea-ice zone around Antarctica. In addition, atmospheric cooling caused by the large areas of glacial ice and sea ice is a worldwide effect.

Croll's speculation was later reinforced by the computations of Milutin Milankovitch, published in 1941. Milankovitch realized that the variations in the tilt (the inclination) of the earth's polar axis with respect to the plane of the earth's orbit around the sun would also change the amount of solar energy received at different latitudes. He took into account both precession and tilt, and was able to calculate the variations in solar energy entering the top of the atmosphere during the summer for the range of Northern Hemisphere latitudes over the last 650,000 years. This date is about half-way back through the Pleistocene, the geologically recent epoch of ice ages. He focused attention on latitude 60°N where major ice-sheet growth begins, although when insolation is high at 60°N it is often high at lower latitudes also. The Antarctic is very cold, and modest changes in temperature there would have little effect on melting. In the Milankovitch hypothesis, significant variation of glaciation takes place only in the

Northern Hemisphere where the larger ice sheets are known to have disappeared during the last deglaciation. The classical Milankovitch idea is that insolation relates directly to glaciation. With each increase of received solar energy near a maximum on a ~10,000 year orbital timescale, the climate warms and the northern glacial ice should melt away. With each decrease toward a minimum, the climate cools and the ice sheets should accumulate and extend their borders south from the nucleation areas. The history of the development of the Croll-Milankovitch idea from its inception to the present has involved efforts to ascertain whether or not orbital effects are the underlying cause of glacial variations, and if not, then how and to what extent orbital changes affect the climate. But the direct-cause hypothesis of warming and cooling dominated the thinking of Milankovitch and that of others since his time.

Initially the glaciations predicted by Milankovitch's calculated valleys and peaks of insolation seemed to be dimly visible in the geological records, and European scientists of the 1930's and 1940's began to favor his hypothesis. However, as more detailed field work was done with more accurate ages to test the hypothesis, contradictions appeared. The new data involved the age measurement of organic remains, such as wood buried by the advances of the ice sheets. The new technique measured the radioactive carbon content of the wood, a method that was developed and described by W. Libby in the 1952 at the University of Chicago and quickly put to use. Radioactive carbon fourteen (^{14}C) is produced in the atmosphere by cosmic rays, and it becomes incorporated in plants and animals by the same biochemical pathways

Post press correction: 14C decays by the emission of high energy electrons, not by emission of helium ions.

followed by stable carbon atoms. Half of any given collection of ^{14}C atoms will decay radioactively in about 5700 years with the emission of high-energy helium ions. By carefully counting the ions emitted from a measured sample of carbon, the decay rate gives a measure of the time since the sample was part of a living animal or plant.

Using the ^{14}C method to measure the ages of tree wood buried in glacial till, it was found that several advances and retreats of the Laurentide ice-sheet front had occurred over the last 80,000 years. This contradicted the simplistic Milankovitch idea that the climate system should smoothly follow the insolation trends. Other embarrassing bits of evidence were found in Europe, and the Milankovitch concept fell out of favor in the minds of most workers in the field. Later in the mid 1960s the pendulum began to swing back in favor of the Milankovitch idea as other different correlations were found in the ocean sediment records. Nevertheless, the case for the direct effect of insolation was weak, and the causal connection between insolation and ice-volume variations remained elusive.

A few scientists remained quite unconvinced of the validity of the Milankovitch hypothesis. Sir Fred Hoyle in his book, *Ice,* in 1981 raised strong objections, based in part on the quite small differences expected for ice-sheet melting due to Milankovitch insolation variations when one considers the high reflectivity of glacial ice and snow. In addition, ice and snow are translucent and are unable to concentrate absorbed solar energy in a thin surface layer for effective melting. It is also true that an ice sheet only melts effectively in an ablation zone at low altitude around the ice-sheet edge. In cold climates this zone will be quite narrow, and would probably not change very much if the only variable were Milankovitch insolation. Hoyle's most relevant argument, however, is that correlations alone cannot "prove" that insolation variation is a primary cause of the ice-volume variation. A good case can be made only if a strong cause-and-effect relation can be shown.

Because of the contradictory details and lack of a convincing cause-effect relation, scientists viewed a connection between orbital insolation changes and ice-age climate variations with great skepticism for many years while the ^{14}C ages of tills were being measured, even as the different marine correlations began to be reported that favored the hypothesis. And a subtle shift in the context occurred. The early thinking had focused on the geographical extent of the ice sheets, but now the new marine data shifted the focus to changes in the world volume of glacial ice, or rather to the proxy in the ocean sediment records from which the change in ice volume could be inferred.

2

Sediment cores from the deep ocean

The midcentury skepticism of the Milankovitch idea began to diminish with the pioneering work on the recovery and analysis of deep-sea sediments initiated by the Lamont Doherty Geological Observatory of Columbia University in the 1950's. The sediment recovery technique was a brilliant success, and a wealth of information was obtained that was immediately related to glaciation and climate. The analysis of sediments by mass spectrometric techniques was also applied with great success by N.J. Shackleton at his laboratory at Cambridge University, and a steady stream of insights on glaciation and climate resulted from these efforts. Indeed, deep-sea sediments have proved to be one of the main keys to the climates of the past few million years.

In the deep ocean, far from river outlets and where the sedimentation rate is uniform and small, an unbroken record of climatic changes affecting the sediments over hundreds of thousands of years can be found in the top few meters of mud on the sea floor. Consequently, where sediment is undisturbed by submarine debris flows or intermittent submarine currents, a valuable climate record can be obtained by drilling techniques. To sample the sediments, a long pipe is driven into the mud below the drilling ship, which is positioned directly above. When the pipe is hauled to the surface, the cylindrical core of undisturbed sediment having the inside diameter of the pipe is extracted. After a preliminary examination on board ship, the sediment core is stored and returned to on-shore facilities where it is archived for future study. In the cores in the North Atlantic from ocean depths above about 4000 m (meters), a large part of the sediment consists of microscopic or pin-head-sized sea shells, the forams of different species of foraminifera (Fig. 2). Some of these species of mollusks live on or in the sediments at the sea bottom (the benthic species) where they feed on the organic debris that rains down from the surface. Others live in a floating mode near the sea surface (the planktonic species) where photosynthesis makes complex food chains possible. After the planktonic foraminifers die their shells sink to

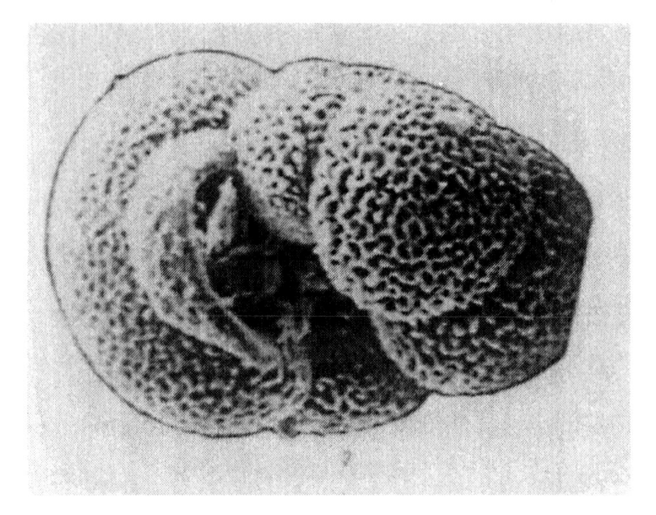

Figure 2: A microscopic mollusk from the Arctic Ocean: *Globigerina paraobesa.* Much of the world's deep-ocean sediment consists of the small shells of foraminifera much like this specimen, which is smaller than a small pinhead. From: "Marine geology and oceanography of the Arctic seas," Y. Herman, ed. Springer-Verlag, New York, 1974. Modified from Plate 11 and used with permission from Springer-Verlag.

the bottom with the benthic shells, and are eventually covered by younger forams and the wind-blown clay and silica dust that continually falls into the oceans.

There are two kinds of analyses of the foraminifera. Analysis of the relative numbers of individuals from each species gives information about past climate temperatures, and analysis of the isotopic composition of their shells gives information on ice-volume changes. Summer sea-surface temperatures are inferred from the percentages of individual shells of each species. Some species live only in cold water typical of the Arctic. Other types prefer warm tropical waters, and still others thrive somewhere in between. By systematically counting the relative abundances of shells of each species in samples of sediment down through the core, the temperature of the surface water at the time the sediment in each layer was deposited can be inferred by comparison with similar modern abundances of species living in environments with known temperatures. From such comparisons, past oceanic climate temperature changes can be inferred.

Obtaining the ice-volume variations from the forams is done by measuring the isotopic composition of the shells of the benthic species, and this is a more complex matter. During the growth of the foraminifer, its shell (its foram) is formed by the precipitation of calcium carbonate ($CaCO_3$) from the carbonate that is in solution in sea water. Variation in the total world volume of glacial ice alters the ratio of the oxygen isotopes in ocean water, which in turn changes the isotopic composition in the foraminiferal shells. The isotopic composition of the shells is simply the ratio of abundances of chemically similar atoms that have different atomic weights. The abundance ratio in the shells varies with the seawater temperature as well as with the sea-water composition, and consequently care must be used in analyzing the isotopic measurements. We are mainly concerned with the relative abundances of the isotopes oxygen sixteen (^{16}O) and the heavier oxygen eighteen (^{18}O). Carbon thirteen (^{13}C) measurements are also used to investigate ocean circulation patterns, but our focus here is on the oxygen isotopes.

To obtain a history of glacial-ice volume variation, individual shells from a single benthic species are collected from each layer of sediment in the core and prepared chemically for analysis in a mass spectrometer to measure the isotopic abundances relative to abundances in an accepted standard carbonate material. The abundance ratios, expressed as $\delta^{18}O$ (delta O 18) are plotted against distance measured down from the top of the core, and, if the temperature of the seawater remained constant and the type of water mass did not change, a record of relative changes in glacial-ice volume over time is obtained.

This technique works because the isotopic composition of the world's seawater changes as large ice sheets grow or shrink. When water from the sea is evaporated, the probability that an H_2O water molecule containing the lighter ^{16}O will evaporate is greater than that of a heavier molecule containing ^{18}O. This is called a fractionation process, in which disproportionately more ^{16}O than ^{18}O is evaporated and transported to the ice sheets and deposited as snow. Therefore, the ratio of ^{18}O in the ocean to an accepted standard, expressed as the oceanic $\delta^{18}O$, becomes larger or more positive when ice sheets are growing, and smaller or more negative when ice sheets are melting. These ocean water changes cause precisely corresponding changes in the $\delta^{18}O$ ratio of the carbonate shells of the foraminifera. The amplitude of the glacial-to-interglacial change in $\delta^{18}O$ of the carbonate is about 1.1‰ (parts per thousand) due to ice-volume change. However, the effect of the temperature change on $\delta^{18}O$ is about 0.25‰ per °C, and consequently temperature changes can also have a large effect on the measured $\delta^{18}O$.

It soon became apparent from the isotope record that many glacial cycles have occurred, as many as nine over the last million years (Fig. 3). Intervals of warm and relatively ice-free climate like today are inferred from the high negative values of $\delta^{18}O$ in the record that occurred approximately every 100,000 years over the last million years. However, it was not obvious how to separate the effect of a change of water temperature from a change of ice volume in the isotope-ratio data. In the earliest pioneering study of C. Emiliani in 1955, it was thought that the main part of the $\delta^{18}O$ variation in planktonic forams (foraminifera living near the sea surface) was due to glacial-climate variations of ocean temperature, and this viewpoint was hotly debated in the scientific literature for a few years. However, the idea became increasingly untenable. Dansgaard and Tauber in 1969 finally showed that most of the variation was an ice-volume effect, although some cores are affected by temperature much more than others.

The $\delta^{18}O$ is relatively free of temperature variations only in the benthic foraminifera that lived near the bottom in the deeper ocean. This is because the deepwater in the world oceans is formed when surface water sinks in polar regions, and it can only sink when it cools to within a few degrees of the freezing point, no matter what the climate in lower latitudes is. Consequently, in the bottom sediments where the benthic forams live, the temperature variations are much smaller. But even there, deepwater masses formed under the Arctic conditions in the Greenland-Norwegian Sea (the Nordic Sea) have a slightly different temperature and $\delta^{18}O$ than those formed in the Antarctic. The benthic records at most sediment locations may be fairly good, but the usual assumption is that benthic temperature changes do not seriously affect the $\delta^{18}O$ proxy for ice volume, and this assumption may not always be correct. With these caveats, the isotope ratios from benthic species of foraminifera were therefore chosen for the reconstruction of the history of glacial ice-sheet volume variation using sediment records that extend back millions of years in long cores from the deep ocean.

If it is to be useful in paleoclimate research, a timescale must be assigned to the sediment record. This is easier to do if the sediment is deposited at a uniform rate over time, and careful efforts are made to ensure that only the most uniformly deposited cores are analyzed. A wide range of deposition rates occurs in the deep-sea, but typical rates are only a few centimeters per thousand years. At first the early drilling technology recovered relatively short cores containing the last few hundred thousand years of sediment history. The timescales were established initially by radioisotope dating techniques using ^{14}C from organic debris, a method that is practical only to about 40,000 years before present (BP). In older sections of uniform cores the ages of the layers in the core could be estimated

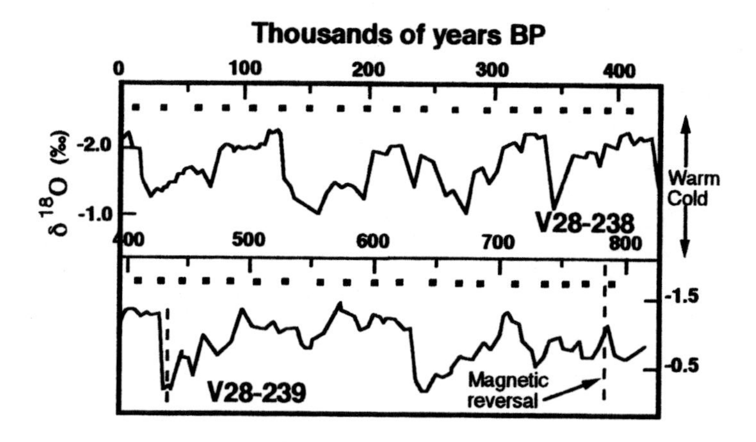

Figure 3: Insolation maxima and an ice-volume proxy record over the last 800,000 years from two sediment cores from the Pacific. Northern Hemisphere insolation maxima at 25°N latitude (square dots) occur approximately every 23,000 years. The benthic temperatures are not identical at the two core sites, giving them somewhat different average isotope ratios. Isotope record from N.J. Shackleton with timescale modified from Johnson (1982), with permission from Academic Press. Square dots, Berger's tabulated data (1978). The upper more negative peaks indicate warm interglacials with minimal glacial ice.

by extrapolating down the core from the calibrated upper levels. Later on, improved ages for occasional points were obtained from volcanic ash layers using other radioisotope schemes. One of the most widely used methods of dating deep-sea core layers today is a correlation method, which uses long cores and involves "tuning," in which major peaks and valleys in the $\delta^{18}O$ record are matched to the astronomically calculated insolation variations. The method depends on the basic assumption that the major $\delta^{18}O$ variations follow insolation changes, a hypothesis that was not widely accepted until 1983 after modern computers enabled more accurate calculations of insolation variations that depend on changes in parameters of the earth's orbit. The method seems to work well most of the time despite the intellectual discomfort that is associated with the basic assumptions that $\delta^{18}O$ changes follow the ice volume and ice-volume changes follow insolation. These assumptions have been made without an acceptable physical mechanism to connect the changes in ice volume to the insolation variations.

3

Accurate calculations of insolation

From the first, the rough correspondence of the major $\delta^{18}O$ peaks in sediments with the maxima of the summer insolation invited attempts to establish rigorous correlations. So there was a need for an accurate knowledge of the orbital parameters of the past from which the Milankovitch variations could be calculated. An accurate knowledge of the insolation variations would enable a precise test of the correlations with the marine oxygen-isotope records. The actual solar energy reaching and warming ice-free land and the ice sheets depends on factors such as the amount of dust, water vapor and clouds in the atmosphere. In the Milankovitch hypothesis the astronomical calculations predict the average of the incoming solar energy incident on the top of the atmosphere over the summer and winter halves of the year, and the modifications of surface temperature due to these other factors are neglected.

There are three astronomical parameters that are needed to calculate the incident solar energy over the summer half of the year at different latitudes. They are the eccentricity (ellipticity) of the earth's orbit, the precession factor giving the position of Earth in its orbit, and the inclination of the earth's polar axis to the plane of the orbit around the sun. All these factors vary due to the changing gravitational effects of the sun, the moon, and the planets. It is therefore a daunting task to accurately compute the three parameters back into the past, even though the masses and orbits of the planets are quite accurately known. Milankovitch and his assistants did this to a limited approximation without the benefit of electronic computers, which was a prodigious effort at that time.

The first major refinement was made by A.D. Vernekar after computers became available. In 1972 he published the calculated global solar insolation at 5° latitude intervals for the past million years, thus providing accurate amplitudes and an accurate time scale for Milankovitch insolation variations. The modern caloric summer average insolation for Northern Hemisphere latitudes is shown in Figure 4.

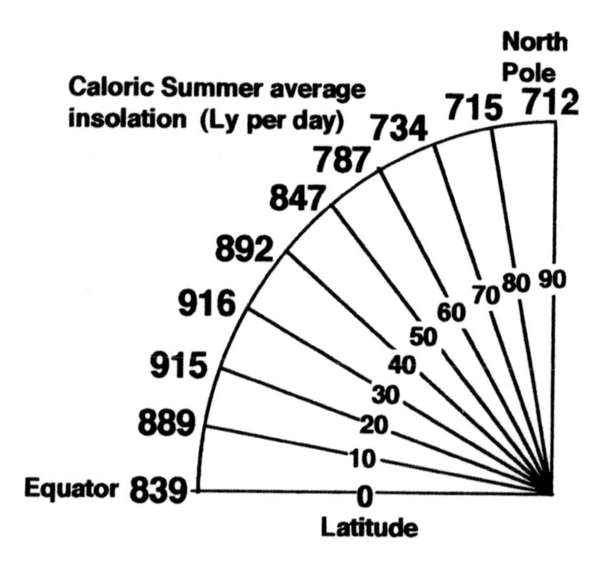

Figure 4: Summer isolation variation with latitude. Today's (1950) average caloric summer insolation in langleys per day. From Vernekar (1972). The caloric summer is defined as that half of the year when every day has more incident insolation at the top of the atmosphere than any day in the other half of the year. One langley is one gram calorie of incident solar energy per square centimeter of surface.

Vernekar's tabulated values are averages that are expressed conventionally as the difference from modern incident insolation at the top of the atmosphere at each latitude, averaged over the summer half-year. It is not difficult to calculate such differences for the longest summer day at the solstice or other intervals such as the three months of summer centered on the solstice day. In 1978 A. Berger published improved results that were obtained by including more terms of the gravitational interaction between Jupiter and Saturn, and he also made the results available in tabulated form. Subsequently, he and colleagues extended the calculations to cover the last five million years. Although the accuracy of the predicted insolation levels decreases for more distant times, the differences between the Vernekar and Berger tabulations are small over the last million years. This suggests that, for the last several hundred thousand years, the accuracy is quite adequate for precise comparisons with marine and other geological records.

These astronomical calculations for the distant past are important because they show that the general nature of the insolation fluctuations is fairly constant, whereas the character of the $\delta^{18}O$ record in sediments has changed greatly over the last three million years. This implies that physical factors in the climate system, such as the North Atlantic Ocean circulation, have altered the way in which Milankovitch insolation variations bring about ice-age climate changes. Indeed, the ocean has a lot to say about ice-age climate variation, in addition to what the $\delta^{18}O$ records of marine sediments tell us.

In addition to the summer sea-surface temperatures

inferred from the abundances of microscopic species of mollusks discussed earlier, the salinity of the sea can sometimes be derived from the measured $\delta^{18}O$ of surface-dwelling species and the inferred temperatures, and salinities often tell us about the aridity of the climate or the effects of large meltwater events. Perhaps even more interesting is the study of deep-sea sediment composition around Antarctica, which reveals the extent of complete year-round coverage of sea ice during the last ice-age maximum, when the permanent sea ice extended much farther out to the north than it does today. We will return to this topic later in the book. But the measurement of ancient sealevels that were determined by the growth of large ice sheets on land was an important step in connecting the orbital changes to glaciation.

4

Glacial-ice volumes and sealevels

We have seen how the variation in $\delta^{18}O$ of benthic foraminifera living in quite deep water can be considered as a proxy for glacial-ice volume changes over millions of years. However, the changes in $\delta^{18}O$ values are not automatically calibrated and can only tell us the relative changes in world glacial-ice volume. They are also subject to errors due to unidentified temperature effects or a variation in the source of deepwater, which could have a corresponding $\delta^{18}O$ variation. World sealevel changes can provide such a calibration because sealevel changes correspond very closely to changes in glacial-ice volume on an ice-age timescale of hundreds of thousands of years.

Each cubic meter of ice, when added as snow to growing ice sheets on land, withdraws an equivalent weight of water from the oceans. This process lowered world sealevels to 120 m below present at the last glacial maximum about 19,000 years ago. This low sealevel has been accurately measured and dated in cores taken from submerged coral reefs offshore at Barbados, as reported by R. Fairbanks of Columbia University in 1990. Barbados is a good place for such research because the Caribbean location minimizes minor departures from the average world-sealevel change when ice sheets melt or accumulate. Different small departures from the average occur at various locations all around the world, because removing water from the world ocean and putting it on ice sheets concentrates its weight on smaller areas and causes global distortions in the geoid shape of the earth that are particularly visible at locations near the ice sheets.

Coral reefs accumulate as corals die and and are broken up by scavengers or wave action. The resulting debris lodges within the dense forest of remaining living corals and slowly becomes cemented into limestone. The corals of interest to us grow in shallow water along tropical coasts, and build up underwater reef-terraces with tops near low-tide sealevel. The reefs therefore record intervals of nearly constant sealevel at extremes of glacial-ice volume variation in the glacial cycle. The terraces are obscured at most stable coastal locations because they lie submerged below the high sealevel of today, and the older ones are also buried within the younger reef deposits. However, such terraces can be viewed separately at locations where moving plates of the earth's crust collide, causing the coast to be tectonically uplifted at a steady rate. If the rate of uplift is sufficiently rapid, a reef formed at one nearly stationary high sealevel will be lifted high above that elevation while the sea falls and rises again to the next high sealevel many thousands of years later. Therefore, reefs that grew at successive sealevel maxima can be found displayed as a series of coral-terrace steps, elevated to various heights along the coast.

Beautiful examples of such terraces that rise from the shore like giant stairs are found on the Huon Peninsula in New Guinea, where coastal uplift rates can exceed two meters per thousand years. On the island of Barbados, where the uplift rate is much smaller, a less spectacular series of terraces occur (Fig. 5), in which the highest on the island is almost a million years old. These sets of terraces represent a long series of high seastands, and the ages of the

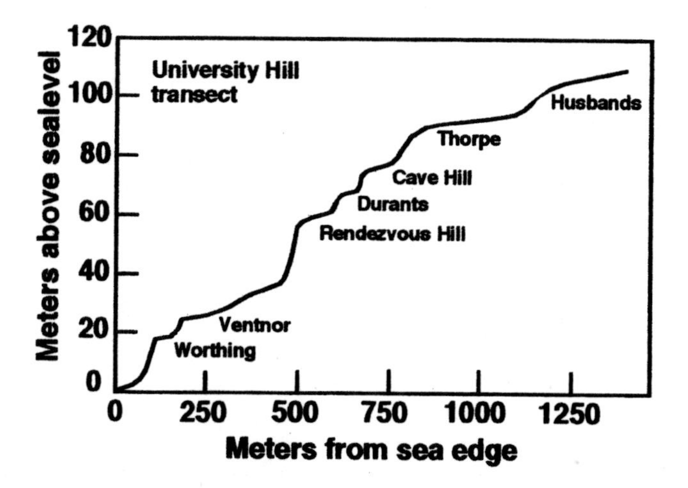

Figure 5: Uplifted coral-reef terraces on Barbados formed in the University Hill neighborhood over the last 350,000 years during intervals of nearly constant high sealevel. The Rendezvous Hill terrace is known as the First High Cliff, and was formed during the last interglacial period, 125,000 years ago. The uplift rate in this neighborhood is about 0.42 meters per thousand years. Maximum terrace elevation on Barbados is 340 m. The transect is from a topographic map in Bender et al. (1979).

terraces can be used to test the Milankovitch proposal that the disappearance of the ice should occur when the insolation is near a maximum.

The terraces on New Guinea were studied early in the 1970s by J. Chappell, and those on Barbados in the late 1960s by teams led by R.K. Matthews. The ages of the terrace corals were measured by the early radioisotope method in which the high energy helium ions from the $^{230}Th/^{234}U$ radioactive decay system are counted. The first measurements reported by J. Chappell (1974) and A. Bloom et al. (1974) using this method clearly correlated several stationary intervals of high sealevel with maxima of

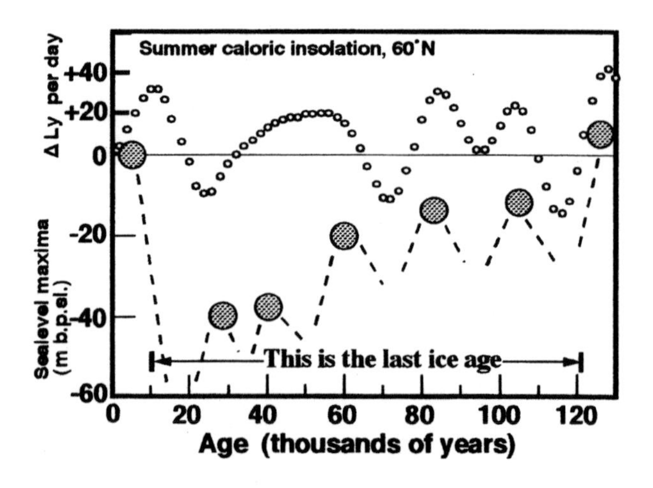

Figure 6: Insolation variation at 60°N and sealevel maxima in the last ice age in New Guinea. Summer insolation differences are relative to 1950 AD. From tabulated insolation supplied by Berger (1978). Maxima of sealevel from Bloom et al. (1974). Used with permission from Academic Press.

high-latitude insolation over the last 130,000 years (Fig. 6).

An interesting aspect of the terraces is that their relative size is an indication of the duration of the interval of high sealevel. Over the last 7000 years of our present warm interglacial period, sealevel has changed very little, the modern coral terraces are quite broad, and where the terraces along the shore have built out into deeper water, the forereef slopes are long and steep. A similar prominent terrace on both New Guinea and Barbados was formed during the last warm interglacial interval about 125,000 years ago. The deep-sea sediment $\delta^{18}O$ records show that this interval lasted about as long as our present interglacial, almost 10,000 years. On New Guinea, however, the stratigraphy of the last interglacial terrace and its approximate age seem to show that the high sealevels accompanying this deglaciation occurred earlier than had been expected from Milankovitch insolation correlations. This disagreement may be the Achilles heel of the conventional Milankovitch hypothesis, as we shall see.

Nevertheless, by 1974 the coincidence of several high seastands and peaks in the $\delta^{18}O$ records with accurately known astronomical ages of solar insolation maxima had greatly strengthened the Milankovitch hypothesis. In subsequent years published correlations continued to accumulate and add support to the idea, although the climate mechanisms that enabled insolation to cause ice-volume changes did not become clear.

An example of the thinking of the time is the Johnson and McClure paper in 1976, which proposed a simple qualitative model for ice sheet variation based on an extension of pack ice and diversion of the North Atlantic Drift caused by lower insolation. This was an unsuccessful attempt to find an additional mechanism to explain the ice-volume variation, because the simple concept of direct domination of the melting rate by insolation remained quite unsatisfactory. For example, 175,000 years ago when the calculated insolation was even greater than that associated with the most recent deglaciation, no transition to a warm interglacial period with high sealevels occurred. At that time, an intermediate-level $\delta^{18}O$ peak appeared (Fig. 3) in the foraminiferal record, and it could be argued that the Milankovitch hypothesis remained intact, but that the melting due to the insolation factor was partly overcome by unknown effects within the earth's climate system. This argument was often repeated whenever departures from ideal Milankovitch correlations were encountered. Such departures, however, were not sufficiently disturbing to invoke a critical re-examination of the fundamental Milankovitch idea that the ice-sheet growth and shrinkage were controlled more or less directly by insolation variations.

5

A heyday for Milankovitch

The decade of the 1970s and early 1980s saw a tremendous acceleration of productive research on the problem of ice-age climate change. The technology improved, and longer deep-sea cores were obtained with records that extended millions of years into the past. The studies of $\delta^{18}O$ variations of marine sediments at the Lamont Doherty Earth Observatory were vigorously pursued, notably under the direction of W.F. Ruddiman. The Lamont teams recovered cores that sampled deep-sea sediments over most of the North Atlantic. Strong programs were likewise established by European groups. The work by N.J. Shackleton at Cambridge University was especially productive. Shackleton and Opdyke (1977) analysed benthic foraminifera from the Lamont core V28-179 and showed that the latest series of ice ages began about 3.1 million years ago when small $\delta^{18}O$ fluctuations due to ice-volume variations first appeared in the record.

These initial variations were not due to a change in the character of Milankovitch insolation, because the calculations by A. Berger show that the insolation fluctuations have not changed fundamentally over the last 5 million years. Then what could have tweaked the Earth's atmosphere-ocean system to start the first oscillations of the ice-age epoch ? The key event was probably the emergence of the Isthmus of Panama, dated to about 3.1 million years ago by L. Keigwin in 1978, and discussed more recently by McNeill et al. in 2000. The emergence established a land bridge between North America and the former island continent of South America (Fig. 7). The age of the emergence is shown by radioisotope dating of ages of marine fossils that had been common to the Caribbean and the neighboring Pacific before the emergence. When Panama emerged, single species living in the sea off the Pacific and Caribbean coasts evolved along different paths, and their fossils became distinguishable as different species.

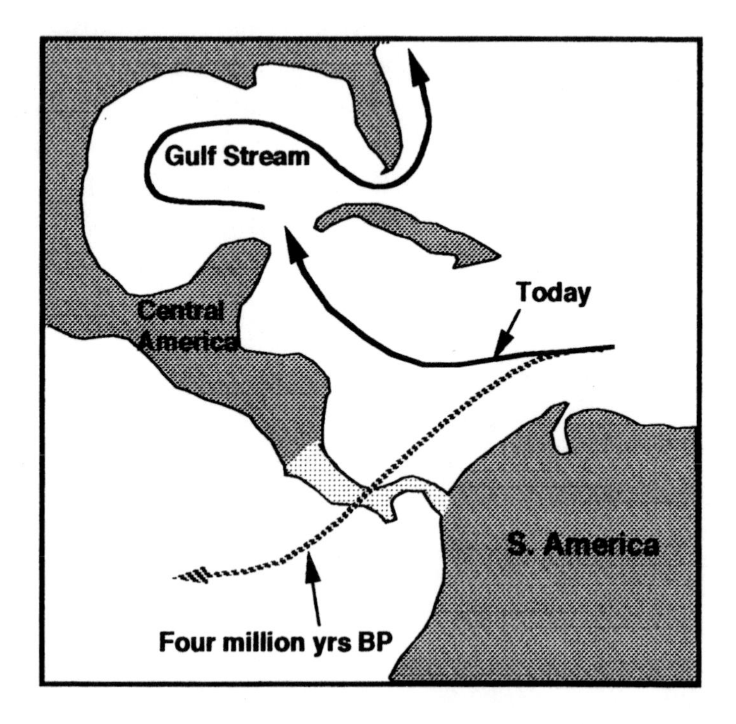

Figure 7: Ocean current change due to uplift of Panama. The isthmus emerged and blocked westward flow about 3.1 million years ago. The emergence diverted warm water into the Gulf of Mexico and increased the warmth of the Gulf Stream and the amounts of atmospheric moisture transported to high latitudes.

The closure of the strait between the continents significantly affected the circulation of the northern North Atlantic. Salty Caribbean water that had previously been blown through the strait by the trade winds was now retained east of Panama. The salinity of the Caribbean increased, and warm salty water that formerly passed through the strait now was added to the Gulf Stream moving northward off the coast of Florida. A greater amount of salty water then reached the seas around Iceland, and started the oceanic conveyor-belt (see Chapter 9), an offshoot of the Gulf Stream. This increased the humidity, the available moisture, and the capability for growth of the small ice sheets of that time in Arctic regions.

One need only consider the modern Greenland Ice Sheet to realize how important a good moisture supply is. The ice cap is 3.4 km (kilometers) thick on central Greenland where snowstorms pass very frequently along and across the coast. Peary Land is located at the northern end of Greenland, 1200 km to the north, far from the main storm tracks, and only about 800 km from the North Pole. There is no significant ice sheet on Peary Land. What little snow falls is melted or evaporated by the brief but strong summer sunlight, and most of that broad area is a polar desert. At the time of Shackleton's analysis of core V28-179, however, most climate scientists placed little emphasis on the importance of an unfailing moisture supply.

For about 0.5 million years after the first oscillations appeared, the amplitudes were small. About 2.6 million years ago the $\delta^{18}O$ amplitudes became larger. The implied larger magnitude of glaciation is consistent with the record of glaciation on land recently reported by Helgason and

Duncan (2001) for Iceland. The isotope ratio periodicities at this time suggest that polar-axis inclination variations were the strongest influence on ice-volume variation. From about a million years ago to the present, the $\delta^{18}O$ oscillations began to show sawtooth cycles of roughly 100,000 years, which contain smaller-amplitude cycles of 23,000 years (Fig. 3) that correspond to the precession effect. The longer cycles were climaxed by inferred high sealevels with warm interglacials like today. Although the visual correlations between $\delta^{18}O$ peaks, dated high seastands, and insolation were intuitively convincing to many researchers, it took a mathematical analysis of high quality benthic $\delta^{18}O$ data and its comparison with the orbital factors to elevate the Milankovitch hypothesis to a respectable status. The breakthrough occurred with a sophisticated analysis of a large amount of carefully researched data by an impressive group of authors.

In this landmark paper published in 1976, J.D. Hays, J. Imbrie, and N.J. Shackleton described the results of a frequency analysis in which cycles of the benthic $\delta^{18}O$ were compared with cycles of the insolation variations. Hays represented the Lamont Doherty powerhouse of activity in the retrieval and study of sediment cores from the deep ocean, Imbrie was an eminent mathematically-inclined geologist from Brown University, and Shackleton at Cambridge University was the leader in mass spectrometric analysis of deep-sea cores. In both the $\delta^{18}O$ and the astronomical data the 23,000 year precessional cycle and the 41,000 year cycle of polar axis inclination variation stood out above all others. This technique of frequency analysis is powerful in that it extracts the principal frequencies from a considerable background of fluctuations due to internal factors that affect the Earth's climate more randomly. The mathematics assigned about 65% of the $\delta^{18}O$ variance to the two astronomical factors, and about 35% to the more random internal factors. This is why the Hays et al. paper was appropriately titled "Variations in the Earth's orbit: pacemaker of the ice ages," because the astronomical factors appeared to regulate the ice oscillations, but did not have absolute dominance.

With the stamp of mathematical respectability on the Milankovitch hypothesis, more scientists began to favor it. The study of climate proxy records for ice volume therefore accelerated, and the Milankovitch concept was applied to many types of investigations related to climate. Deposits of sediment in lakes or inland seas, which were sensitive to evaporation and precipitation changes associated with high Milankovitch summer insolation many millions of years ago, were found to consist of layer thicknesses having the 23,000 year precessional periodicity. In another application, astronomical correlations were used to measure the age of the last reversal of Earth's magnetic field. The identification of precessional insolation peaks and valleys with features in the benthic $\delta^{18}O$ record was extended by R.G. Johnson in 1982 to a prominent Northern Hemisphere insolation peak that occurred 788,000 years ago, near the last major magnetic reversal found in marine sediments (Fig. 3). Over such a long interval of time, sedimentation rates are never strictly uniform and corrections for slight changes in deposition rate were necessary. This was done by successively aligning the deepest valleys in the $\delta^{18}O$ record with the nearest deep minimum point on the insolation timescale. The deep valleys in the $\delta^{18}O$ occurred at approximately 100,000 year intervals, and the alignment was justified by analogy with the last glacial period in which the glacial maximum occurred during an extreme insolation minimum. In the final analysis of the $\delta^{18}O$ data in a note added in press on page 145 of the Johnson paper, the age of the magnetic reversal was placed at about 783,000 yr BP.

This age, obtained by the new and unconventional method, was considerably older than the previously accepted age of 730,000 yr BP (Opdyke, 1972). New methods are not always welcomed and the result was not widely accepted until 1990 when N.J. Shackleton and colleagues reported a similar result obtained by a more detailed analysis. A short time later an almost identical age was obtained by independent radioisotope methods. Johnson's realignment of the $\delta^{18}O$ minima at six points on the timescale between 350,000 and 800,000 years ago was an example of tuning the marine record to the astronomical timescale, a technique that was widely used in later years. In 1987 Martinson and his colleagues refined the correlations of Pisias et al., making use of the tuning technique to establish a widely-accepted high-resolution chronostratigraphy back to 300,000 yr BP (Fig. 8), now known as the SPECMAP marine sediment timescale.

In the 1970s the work at Lamont and other centers for marine sediment studies was formally organized into an international study of the worldwide climate conditions at the last glacial maximum. The CLIMAP study (Climate: Long-range Investigation Mapping and Prediction) of world ocean surface conditions was a model of cooperation in a

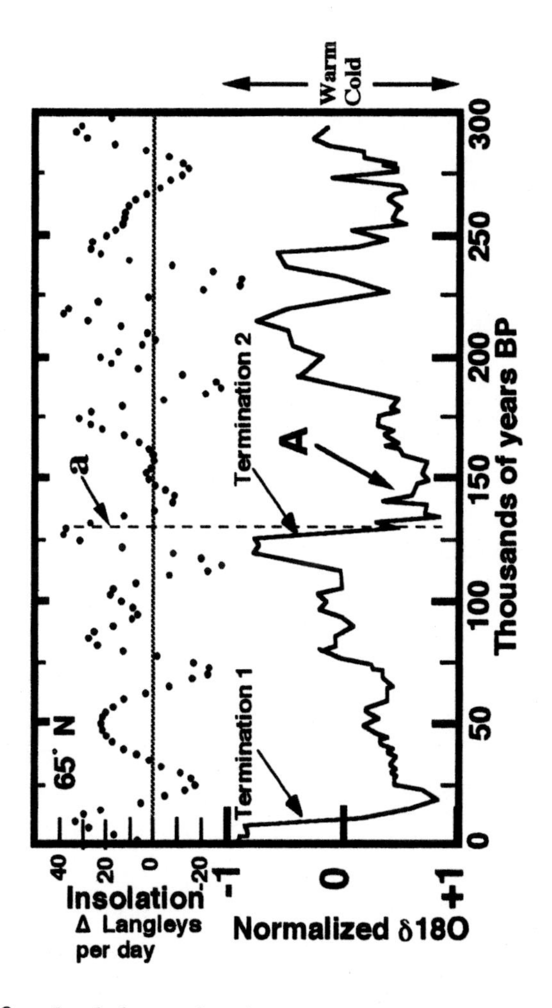

Figure 8: Insolation and a long proxy record of ice volume. Insolation at 65°N compared with the oxygen isotope record on the SPECMAP timescale of Martinson et al. (1987). Used with permission from Academic Press. Tabulated values supplied by Berger (1978). Terminations 1 and 2 resulted in deglaciations that warmed the deeper depths of the North Atlantic. The anomalous deglaciation at *A* occurred with colder benthic temperatures, see Chapter 11. *a* marks a rise to high insolation during anomalous glacial accumulation.

scientific endeavor, and used quite sophisticated methods of foraminiferal species' population abundance analyses to derive the summer sea-surface temperatures at locations around the world at the last ice age maximum.

In nearly all of these studies of the $\delta^{18}O$ in marine sediments, the data appeared to be consistent with the Milankovitch hypothesis. This accumulation of evidence solidified the opinions of the majority of workers in the field that Milankovitch insolation did indeed exert a dominant control on the volume of glacial ice and so also on the ice-age climate. The history of the development and general acceptance of the Milankovitch hypothesis was told in great detail in the book: *Ice Ages: solving the mystery*, written by J. Imbrie and K.P. Imbrie. This 1979 book has probably done more than any other to encourage public awareness of our ice-age heritage.

In the scientific community, interest in the orbital influence on climate change reached a climax at an international workshop held in rural Virginia in May of 1983 and sponsored jointly by NATO and Columbia University. The focus of the workshop was on the evidence supporting the Milankovitch hypothesis. A large number of papers were presented that reviewed the marine and land evidence and the results of computer models, nearly all of which were consistent with Milankovitch. The frequency analyses of $\delta^{18}O$ data from marine cores and from orbital insolation variations took center stage. In the final session of the workshop, representatives from the media and scientific journals were present, with reporters occupying all the front row of seats in the auditorium. With one exception, the content of the afternoon's papers delivered a final and convincing verdict, making the Milankovitch hypothesis the conventional wisdom of the scientific community, with the verdict officially conveyed to the world at large by the media representatives. In that single exception, R.G. Johnson described the high-sealevel coral record from New Guinea, reported in 1974 by J. Chappell. In Chappell's paper a pair of last-interglacial high sealevels was reported, one of which was higher than today. The earlier peak was associated with corals that had been dated to about 135,000 yr BP. Johnson pointed out that this implied a major deglaciation during the previous 10,000 years during a Milankovitch insolation minimum. A heated discussion followed, and this evidence was quickly discounted by an authoritative opinion that the New Guinea corals must have been of poor quality and contaminated, and so gave a wrong age. John Chappell was not present to defend his data.

As we shall see in Chapter 7, the ages for these corals were reasonably accurate, and the assumed Milankovitch correlations between insolation and ice volume did in fact fail completely between 144,000 and 130,000 yr BP.

6
Doubts

Why was the Milankovitch hypothesis of thermal control of ice-age glaciation by insolation changes so widely accepted ? There are, of course, the many published correlations of $\delta^{18}O$ variations with the calculated insolation variations on the astronomical timescale. There are the dated high sealevels with ages that fall close to insolation maxima. And there is the correct measurement of the age of the last reversal of the earth's magnetic field that was accomplished by using the insolation-$\delta^{18}O$ correlations. All these pieces of evidence were glued together by a subtle process involving intuition. Underlying the Milankovitch thinking is the intuitive idea that when northern summer sunlight is high, ice sheets would naturally melt away, and when it is low, they naturally should grow. But things are not that simple.

It has long been recognized that ice-sheet growth often resumes while insolation is still high. The sealevel fall after 82,000 yr BP is a good example. A commonly offered explanation is that the ice sheets themselves cool the climate, and this positive feedback tends to make them grow, even when northern insolation is not low. But how then, can one explain the many contrary occasions when, with less insolation and greater ice volume, deglaciation occurs ? The factors in the earth's internal climate system (excluding insolation) must be quite decisive at such times. In spite of that perplexing thought, the Milankovitch idea has had an almost hypnotic appeal that has been maintained by continuing studies of the correlations between proxies of ice volume and insolation variations.

The last deglaciation illustrates the best known of such correlations. Our collective intuitions were seduced by the early recognition that the northern ice sheets began to melt away soon after Northern Hemisphere Milankovitch insolation began to rise from the minimum that occurred shortly before 20,000 yr BP. At the insolation maximum about 10,000 yr BP the ice sheets were in rapid retreat, and by 6000 yr BP while insolation was still quite high, the last of the major northern continental ice sheets had vanished. The Fairbanks curve of rising sealevel as the insolation peaked is shown in Figure 9. It is ironic that if the scientific community had existed and had performed its studies near the end of the last warm interglacial period 121,000 years ago, a time much like the present, a quite different picture of the previous 20,000 years of ice-volume change would have emerged, and it is unlikely that the Milankovitch concept would have been so easily accepted.

Although the correlations published over the last three decades of the 20th century were not perfect, leading to nagging questions, these imperfections could be assigned to the 35% of the variance that the frequency analysis failed to explain in 1976. And, so it was thought, there were no examples of diametrically opposed trends - when the ice went away during an insolation minimum or when the ice sheets advanced during periods of high insolation. But now such heretical examples have been found. These examples will be described in Chapter 7. To set the scene, let's consider some of the embarrassing but less spectacular examples of inconsistencies with the Croll-Milankovitch

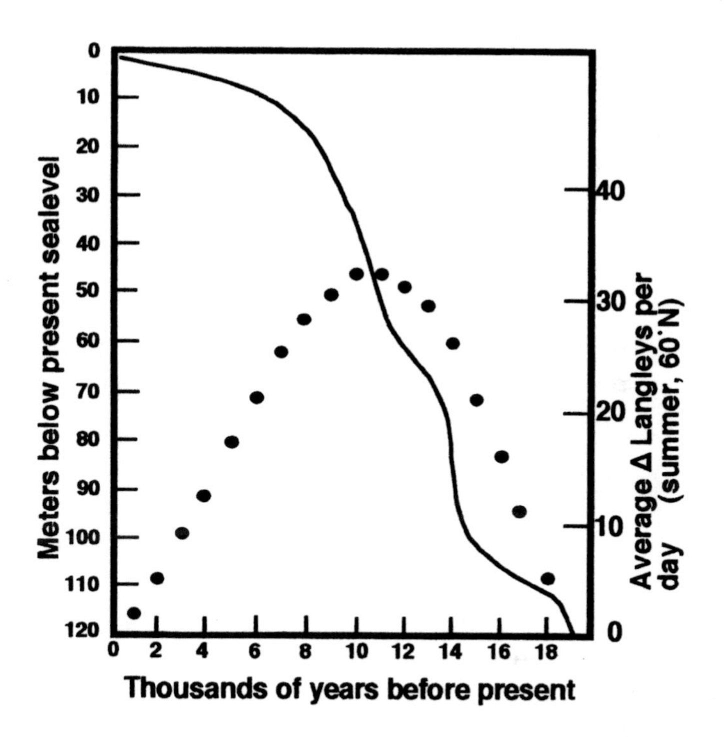

Figure 9: Sealevel rise and insolation during the last deglaciation. Caloric summer insolation at 60°N latitude. Sealevel-rise curve of the last deglaciation modified from Fairbanks (1990), based on offshore core data from Barbados. Tabulated values from Berger (1978).

idea. And it doesn't hurt to recall Fred Hoyle's admonition that correlations can not demonstrate a primary cause.

It has been known for some time that the sealevel record of dated reefs and beaches did not strictly follow the fluctuations of sealevel inferred from the marine sediment record. In 1986 Chappell and Shackleton showed that world sealevels indicated by reefs on New Guinea late in the last glaciation were 40-50 m below today, whereas levels inferred from the benthic $\delta^{18}O$ variations of core V19-30 were almost 90 m below today. The implication is that water in the deep ocean had become 1.5 °C colder at that deep-sea core site then than it is today. The colder temperature would be consistent with an increasing dominance of the deepest parts of the world ocean by the deepwater that forms when colder dense surface water sinks in the Antarctic. That deepwater is a few degrees centigrade colder than deepwater formed in the more saline high-latitude North Atlantic, and if the contribution of North Atlantic deepwater to the world ocean diminishes as the ice age progresses, the benthic mixtures of water all over the world would cool slightly. If the northern deepwater production were to decrease suddenly, the new lower temperature of the world ocean would be achieved in about a thousand years, the accepted mixing time of the oceans. Unless a correction is made for colder benthic water, the ice volume inferred from the $\delta^{18}O$ would be too large, and so would disagree with the New Guinea sealevels. This is certainly the explanation, and it emphasizes the vulnerability of the customary assumption of constant benthic temperatures in the interpretation of the $\delta^{18}O$ proxy record of glacial-ice volume.

A controversy has accompanied persistent hints that the Milankovitch hypothesis did not work very well just before the last warm interglacial period (Fig. 10). In northern Europe, the last interglacial interval, known as the Eemian, was unusually warm. It began about 126,000 years ago, with the first indications of a warming trend about 2000 years earlier. The controversy has focused on

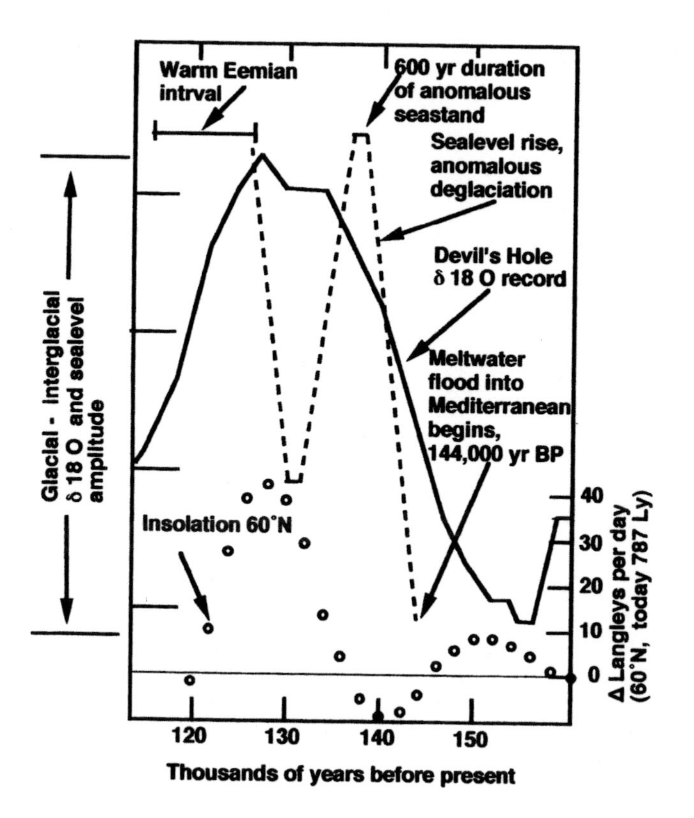

Figure 10: Confusing records of the last interglacial period. The warm Eemian interval applies to Western Europe and the Devil's Hole record (Winograd et al., 1992) to the southwestern United States. The anomalous sealevel rise is from Johnson (2001) and subsequent fall from Esat et al. (1999). Insolation from Berger, 1978 tabulation.

the length of the warm interglacial interval and the duration of the interglacial high sealevel of that time. In the marine records the indicated duration is only 5000-8000 years, depending on just where one defines the top of the $\delta^{18}O$ peak. The Eemian in Europe occurred between 126,000 and 116,000 yr BP. The maximum Eemian warmth occurred slightly later than the insolation maximum. This is consistent with conventional Milankovitch thinking, and there was no visible interruption of high sealevel in northern Europe during that time (Fig. 10). However, other data suggest interruptions, with two or three distinct intervals of closely spaced high sealevel. In the Mediterranean on the island of Mallorca, three last interglacial beach deposits above present sealevel have been reported by P. Hearty (1987), two of which are separated by a surface that was strongly weathered by long exposure to the atmosphere. Two seastands are also recognized in the Bahamas by Neumann and Hearty (1996), in the Carolinas by Hollin and Hearty (1990), and by other workers in many other parts of the world.

A longer interval of generally high sealevels resulting from an early deglaciation is consistent with a long warm interglacial inferred by I. J. Winograd and his colleagues (1997) from the Devil's Hole isotope ratio record in travertine from a deep spring-hole in the southwestern United States. The $\delta^{18}O$ values show an interval of almost 20,000 years of warm climate. The thorium-uranium radioisotope ages for the warm interval, obtained from these trace elements in the travertine, extend from 140,000 to about 120,000 yr BP, which is twice the duration of the warm climate in northern Europe (Fig. 10), and is inconsistent with the Milankovitch hypothesis.

These many scattered pieces of evidence that seemed to be inconsistent with classic Milankovitch ideas have not discouraged persistent efforts to find favorable correlations that support Milankovitch. However, as noted earlier, correlation does not necessarily imply cause, and in a rare critique, Karner and Muller (2000) recently discussed the striking lack of causal connections in the Milankovitch framework. But no better model of climate change had been proposed and very little unambiguous or unchallenged contradicting evidence had been found.

Let's consider again that long-standing puzzle of the last interglacial high seastands on the Huon Peninsula in New Guinea, where the embarrassing age of the first high seastand between 130,000 and 140,000 yr BP implied a deglaciation during an insolation minimum. Opposed to this implication is the lack of a large negative rise in the $\delta^{18}O$ curve at this time on the marine isotope timescale of Martinson et al. (Figure 8). A rise of about 1.1‰ would have been expected from the deglacial meltwater entering the ocean. No significant rise occurred, and this was viewed by many as an absolute bar to the acceptance of the validity of Chappell's 1974 data. Except for minor departures from the ideal behavior of benthic $\delta^{18}O$, variation of the $\delta^{18}O$ was generally believed to be a faithful proxy for variation of the volume of glacial ice and the corresponding world sealevel.

The importance of the possible failure of the $\delta^{18}O$ to follow ice-volume change at that time and the possible contradiction of the Milankovitch hypothesis raised by Chappell's data was not totally ignored, but subsequent expeditions to New Guinea did not resolve the age and stratigraphic questions there. There are two main reasons for this. First, there are uncertainties in the coral ages because it is difficult to find exposed fossil corals in a wet tropical environment that are free of changes in the proportions of the radioisotope trace elements needed for the age measurements. This can lead to large errors in the measured ages. Only at unusual locations where the climate is dry and the corals are never exposed to groundwater and seldom to rainwater is the fossil coral quality sufficiently good that accurate ages can be obtained. Second, many of species of corals live within a range of water depths from 0-20 m or more, and unless carefully selected, the sample fossil corals may not be close to the stratigraphic location on the reef where sealevel was at its maximum. So even if the measured coral age is true, it may not be the age of maximum sealevel and it will probably not indicate the elevation of the maximum. The measured age is more likely to be a later age than that of the maximum because the corals of earlier ages are probably buried deep within the reef by later deposits. Therefore, although Chappell's age of the early anomalous seastand on New Guinea was probably not far off the mark, stratigraphic uncertainty and the age uncertainty have cast a shadow of doubt on the New Guinea results, and have prevented the acceptance of an anomalous high seastand that contradicts the Milankovitch hypothesis, based on Chappell's data. As we shall see, however, thirty years after Chappell's work, new data from the Huon Peninsula reveal a quite different and undeniable contradiction.

7

Decisive contradictions

In 1990 R.G. Johnson retired from a career in industry and joined the Department of Geology and Geophysics at the University of Minnesota to pursue a long-standing interest in the causes of climate change. In considering possible ways to attack the perplexing problems surrounding the last warm interglacial climate interval around 125,000 yr BP, it was noted that, although work on New Guinea had continued intermittently, no further studies of the last interglacial formations on Barbados had been done in recent years. Perhaps Barbados seemed less promising to earlier workers because its rate of coastal uplift is less than one fifth of typical rates on the Huon Peninsula in New Guinea. Consequently, the uplifted terraces are less clearly exposed, and it probably seemed easier to decipher the puzzling stratigraphy by going to New Guinea where, however, the anomalous formation was somewhat ambiguous. However, one of the stated reasons for not accepting the validity of the anomalous seastand on New Guinea was that no similar seastand terrace had been reported on Barbados. An aphorism in geology is: "Absence of evidence is not evidence of absence." Although old and hackneyed, this statement applied perfectly to the Barbados terrace. No one had reported the anomalous terrace there because apparently no one had ever looked carefully for it. Uplift rates are low, and it was not a notable feature of the landscape.

After getting settled in an office in Pillsbury Hall, the

classic old sandstone building that had housed the Geology Department for over a hundred years, Johnson's first objective became an expedition to search for the anomalous terrace of the controversial last interglacial period on Barbados (Fig. 11). A briefing by R.K. Matthews, a veteran geologist who had done much of the early work on Barbados, and some reading of the published literature were quite helpful in getting oriented. Barbados was settled by the first English colonists in the 1620's and has evolved into a densely populated, modern and independent country with strong British traditions and culture. The residents are friendly and hospitable, and with the sun and tropical warmth, tourism is a major factor in the economy. Geologically speaking, it is young, having begun to emerge from the sea about a million years ago as a consequence of a minor collision of plates of the earth's crust that make up the seafloor. The ancient seafloor sediments that are being pushed upward from beneath are capped by a thick layer of coral limestone over most of the island. The limestone accumulates from the debris of underwater coral forests, growing in the shallow-water depth zone that fringes the slowly rising coasts.

Johnson's first expedition to Barbados in 1991 was funded by a National Geographic Society grant, and consisted of Johnson, R.L. Edwards, and his graduate student, C.D. Gallup. Edwards had recently developed a greatly superior thermal ionization method for mass spectrometric dating of coral ages (Edwards et al., 1986/87) using trace amounts of radioisotopes, a technique that might be used to accurately date the age of the anomalous terrace. The objective was to collect fossil corals from both last interglacial terraces, and to measure and compare their

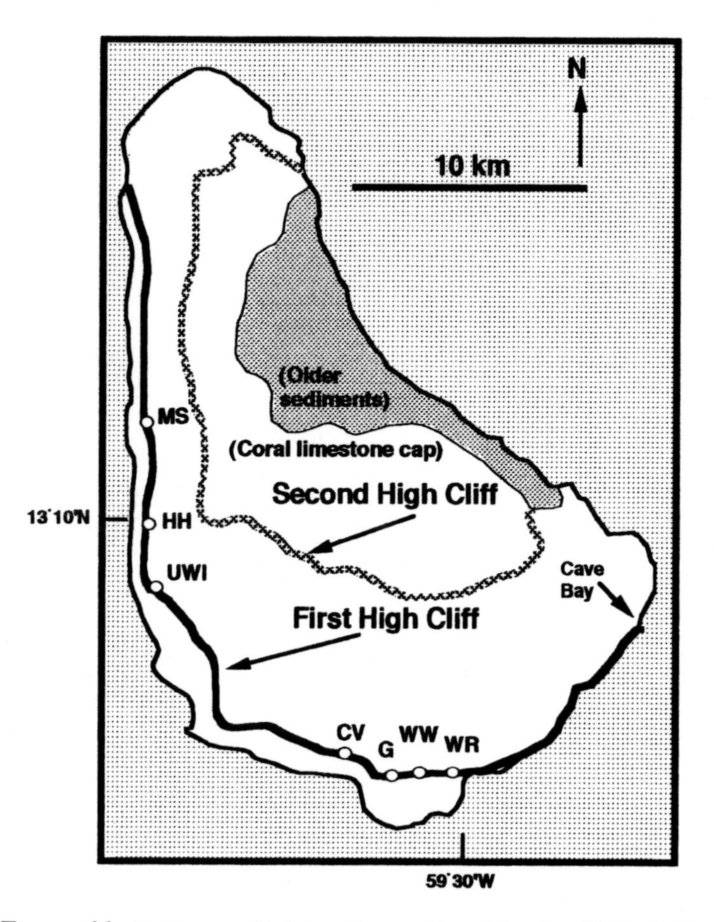

Figure 11: Features of interest on the island of Barbados. Neighborhood locations on the First High Cliff were surveyed to provide differential uplift data for the determination of the age of the anomalous deglaciation. MS - Mount Standfast, HH - Holders Hill, UWI - University of the West Indies, CV - Cane Vale, G - Gibbons, WW - Wilcox West, WR - Wilcox Ridge.

ages. The expedition was a learning experience, as first visits to an unfamiliar field area usually are. Corals were collected on the south and west coasts where uplift rates vary between 0.15 and 0.42 m per thousand years, and where fresh coral exposures occur in recent highway and street roadcuts. To find high quality fossil corals with chemistry unchanged by rainwater, it is necessary to collect them from roadcuts on hillsides. There rainwater runs off without penetrating the hard limestone surface, and corals in the roadcuts below the surface have only recently been exposed to the air. The anomalous seastand was visible as a modest terrace step slightly higher than the maximum point of the main seastand (Fig. 12). Coral samples were collected from several reef formations including the main last interglacial terrace. Unfortunately, the roadcuts through the outer crest of the main terrace did not extend back into the older anomalous coral reef. Therefore no corals were collected from it. Although the coral collection was somewhat disappointing, the expedition itself was not. Spend the day working on the terraces in the hot sun, take a swim on the white sand beach by the hotel at sunset, and enjoy a superb evening meal at any one of many fine restaurants. Few geoscientists enjoy better field conditions than Barbados offers. And as a result of the expedition, a correspondence with W.F. Precht proved to be a key turning point in the Barbados study.

Precht, who was also interested in Caribbean geology, cited a paper by Herweijer and Focke (1978) describing the evidence for last interglacial sealevels on the islands of the Netherlands Antilles. These desert islands are not uplifted. Nevertheless, the last interglacial coral formations depicted there were all somewhat above present sealevel. In their schematic of the stratigraphy of the islands, they showed terraces formed during two major seastands, and one minor

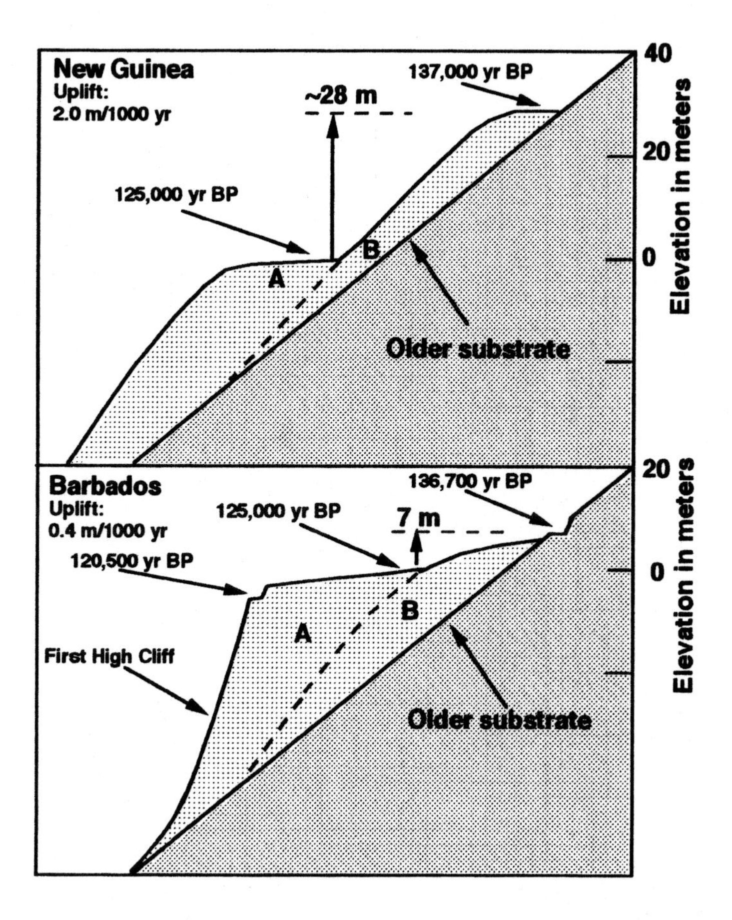

Figure 12: Barbados and New Guinea terraces compared. Schematic comparison of the fossil coral-reef terraces formed during the last interglacial period. The elevation zero point is the sealevel maximum during the warm Eemian period in western Europe, about 125,000 yr BP. On New Guinea, where uplift rates can be 2 m per thousand years, the terraces A and B can be seen as distinct step-like formations. On Barbados, where the uplift rates are small, they are hardly distinguishable at most locations. Earlier workers had not recognized the higher terrace on New Guinea as having been formed early in the last interglaciation.

high seastand lying slightly above present sealevel. These seastands had associated notches eroded by wave action into older coral limestone. In considering the likely sequence of seastand events shown by the Antilles coral formations, Johnson noted that on Barbados the elevations between the terrace components (Fig. 13) would be different at different locations along the coast, depending on the rate of uplift at each location. Johnson had a mild epiphany when it was realized that such elevation

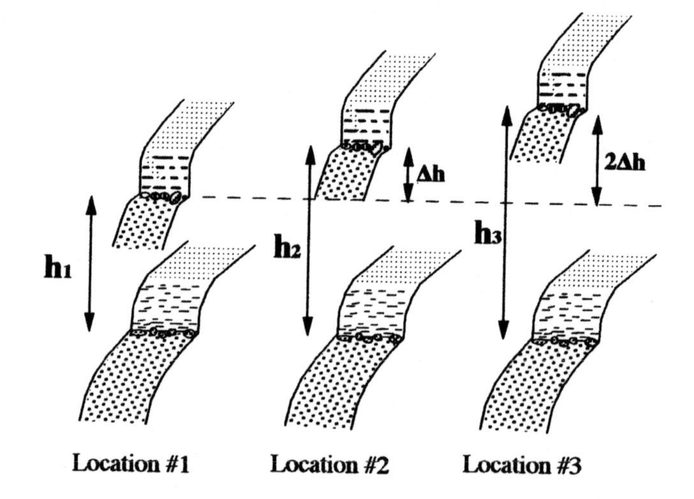

Figure 13: Uplift differences for wavecut notches. Schematic effect of different uplift rates on the elevations of notches made at various locations. The notches are depicted at the time when the lower notch was formed. At location #1 no uplift has occurred. At location #2 an uplift of Δh has occurred since the upper notch was formed. At location #3 the uplift rate is twice as fast, resulting in an uplift of $2\Delta h$. The elevation differences are "frozen in" while subsequent uplift carries the notch pairs to different higher elevations on the coast.

differences between the oldest and youngest member of each notch pair at various coastal locations could enable a calculation of the age differences between the two terraces. If such pairs of notches could be found and elevation differences measured, this new differential uplift method could resolve the controversy surrounding the age of the early seastand on New Guinea.

Where the last interglacial terraces had formed on the slowly rising coastline, one could predict that, under favorable conditions of water depth and wave approach, there would be small notches or dents that were cut by wave action into the hillside at the low-tide level when the sealevel was precisely at its maximum. These precisely defined notches are small, but they do exist at elevated locations along the coast where the ancient water depth was right and wave action was favorable. The first year of exploratory work was followed by six more annual visits to Barbados in which a total of seven suitable notch pairs were located and surveyed (Fig. 14). The elevation differences between members of each notch pair were carefully measured with a theodolite, and the elevation above sealevel of the youngest notch in each pair was determined. From the analysis of this data the age difference between the youngest and oldest seastands was obtained. The result throws a sharp light on the controversial duration of the "last interglacial," a term that now seems somewhat ambiguous.

It requires only a modest effort to consider the theory of the differential uplift method. (If theory is not your thing, just skip ahead a couple of paragraphs.) If the uplift rate at each point along the coast is steady, the elevation

Figure 14. Surveying at the upper notch at the University of the West Indies site on Barbados. R.E. Higashi assisted in the survey on this occasion and is shown standing on the notch formed in an older forereef by wave erosion during the early anomalous sealevel maximum 136,700 years ago. The white coating above the notch is a thick film of carbonate precipitated out on the wet substrate between ancient high and low tide levels. Photo by R.G. Johnson.

difference between any two notches at the same site is given by an equation of a form that is familiar in elementary mathematics:

$$h = ho + (\Delta age)\, R$$

where h is the measured elevation difference, ho is the difference between world sealevels at the time each notch was formed, Δage is their age difference, and R is the uplift rate at the site. The uplift rates were estimated from the age of the youngest notches and their elevations above sealevel. The elevations above sealevel were found by surveying to government elevation monuments and to reliable contour points on topographic maps. The notches were being eroded into the edge of the last interglacial terrace by wave action right up to the time of sealevel fall when the ice age began. Their age was found by correlation with stable seastand formations in the Bahamas, western Australia, and Bermuda, all being formed at a world low-tide sealevel of + 3.0 m at the time the sealevel fall started. In the formations in Australia and the Bahamas the minimum ages among 32 reliably dated fossil corals that grew quite near the top of coral reefs at the end of the last interglacial showed that coral growth stopped at 120,500 yr BP due to the sudden onset of falling sealevel (Chapter 12).

The beauty of this equation is that it provides a way to measure the age difference between the oldest and youngest high seastands without having to find suitable corals from either of the notches themselves for dating. The age and elevation of the youngest seastand above present sealevel is known reliably from Australia and the Bahamas, and this age, together with the elevations above sealevel of the youngest notches on Barbados, enables the calculation of the R values at each location. With all the data in hand, the h values were plotted as a function of the R values. The resulting plot is shown in Figure 15.

The slope of the best analytical straight line through the points is the age difference between the two seastands, and the y-axis intercept of the straight line is the *ho* value, which is the difference between the world sealevels when the members of each notch pair were formed. The resulting age difference was 16,200 years and the *ho* is 4.4 m. Adding the 120,500 yr BP age to the age difference of 16,200 years gives the age of 136,700 yr BP for the point in time when the sealevel began to fall from the early

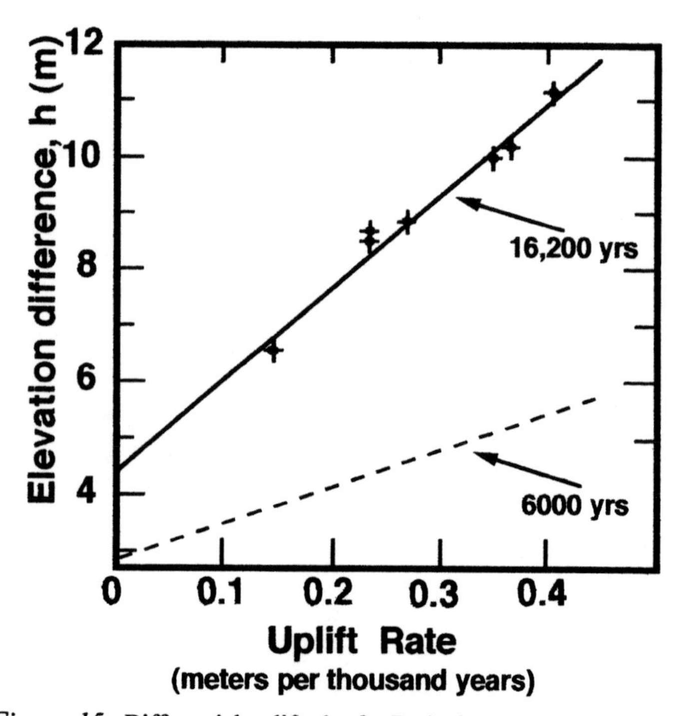

Figure 15: Differential uplift plot for Barbados notches. Comparison of regression lines on plots of elevation differences versus uplift rates for the anomalous sealevel maximum (solid line) that is 16,200 years older than the third notch, and the second sealevel maximum (dashed line, no points) that is only about 6000 years older than the third notch, which was formed just before the last glaciation began and sealevel fell. Upper regression data from Johnson (2001). Used with permission from The American Geophysical Union.

anomalous high seastand that occurred when both insolation and $\delta^{18}O$ values were quite low. The resulting height of the anomalous seastand is 3.0 + 4.4 = 7.4 m above present sealevel. The sealevels of the so-called "last interglacial" on Barbados, including the anomalous seastand, are depicted in Figure 16.

There are strong advantages of this method in the

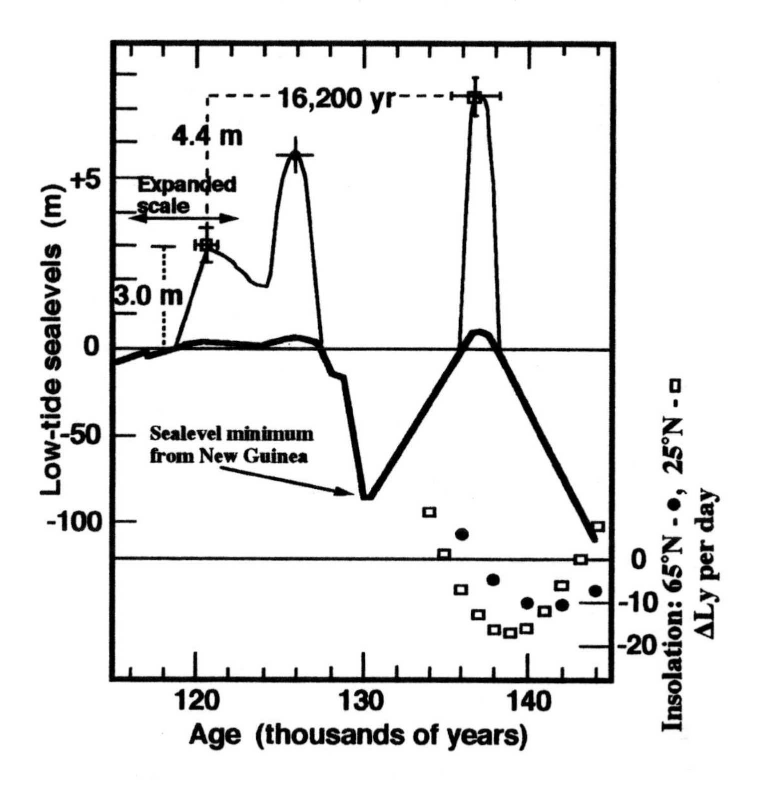

Figure 16: An anomalous high seastand on Barbados. The deglaciation from 144,000 to 137,000 yr BP that resulted in the +7.4 m sealevel was contrary to Milankovitch because it occurred entirely during an interval of low insolation. Modified from Johnson (2001) with permission from the American Geophysical Union. New Guinea sealevel minimum from Esat et al.(1999).

context of the Barbados terraces when compared with the conventional method of dated coral stratigraphy: (1) The notch steps are a precise indication of the low-tide elevation at the end of each stationary interval of maximum sealevel. The accuracy of the measurement of notch elevation differences is within ±0.2 m. This is an accuracy that is impossible to obtain with corals from tectonically uplifted locations because they may have grown many meters below the maximum elevation of sealevel and at a later time. (2) The good straight line formed by data points in Figure 15 implies that the uplift rate has been uniform at each surveyed site, and also implies high accuracy and high precision for the age difference between oldest and youngest notches. Radioisotope ages of corals usually lack the accuracy to enable such a precise age difference measurement. (3) The uplift rates are so small on Barbados that there is no significant error in correlating the maximum elevation of world sealevel with the flat step of the notch. On New Guinea where uplift rates are much higher, some error could be introduced using the differential uplift method.

On the Barbados terraces that were surveyed, the slopes are well drained, and erosion of the surface over the last 137,000 years is negligible. Rain water causes minute dissolution of the coral debris that cements and seals the top layers of the surface as soon as sealevel falls. The surface water flowing down the slope is nearly saturated with carbonate from above. Consequently, the skin of new carbonate several millimeters thick that precipitated out on the old coral substrate in the zone between high and low tide during the 137,000 year-old seastand is still visible today at the UWI, Cane Vale, and Wilcox Ridge sites. Comparable preservation at all the sites surveyed makes the notch elevation differences a true measure of the uplift-modified differences between the ancient sealevel maxima.

Therefore, the age of 136,700 yr BP for the end of the anomalous high seastand is a clear-cut and reliable result, and is consistent with a large amount of stratigraphic data suggesting an early high sealevel. One might think that the journal paper (Johnson, 2001) that eventually reported the result, when under review, would have been accepted by all peer reviewers, though perhaps with some reluctance. Not so. The Milankovitch concept and the faith in the reliability of the $\delta^{18}O$ proxy for ice volume had become deeply embedded in the thinking of the time, and one unnamed reviewer objected, saying: "The marine $\delta^{18}O$ must follow sealevel !" As a wise man once said: "Sometimes nothing is more difficult than to replace an idea with a fact." There is an epistemological problem here. After all, what we "know" is what we are convinced of, and conviction can be stronger than data, stronger than facts.

Consider the implications: (1) The 136,700 yr BP age implies a major deglaciation that was completed during a minimum in Milankovitch insolation, which was a heresy to many reviewers. (2) Assuming the correctness of the accepted marine SPECMAP timescale of Martinson et al. (Fig. 8), the lack of a large negative rise in the $\delta^{18}O$ implies that the effect of deglaciation on the $\delta^{18}O$ values was masked by some unknown factor. The expected negative rise in the $\delta^{18}O$ values was about 1.1‰ for a sealevel maximum of +7.4 m even without a normal benthic warming, but the observed $\delta^{18}O$ rise was hardly 0.2‰. Could the SPECMAP timescale be wrong ? That timescale is almost certainly correct because, although it is "tuned", the longer tuned timescale yields the same precise age of 783,000 yr BP for the last magnetic reversal event as that measured quite independently by radioisotope methods. A reviewer with unlimited faith in the $\delta^{18}O$ proxy for ice volume would indeed face a puzzling decision when viewing the conflicting Barbados evidence.

This conflict can only be resolved by challenging the basic assumption, discussed earlier, that there is never any large benthic temperature change affecting the measured $\delta^{18}O$ values. The assumption of constant benthic temperature is therefore in question, and we are quickly led to the argument that this assumption failed completely during the anomalous deglaciation, and benthic temperatures at the depths of most cores became 3-4°C colder. Most cores are recovered from deepwater where water temperatures today are 4°C or higher, but the greatest depths in the North Atlantic are occupied mostly by Antarctic Bottom Water at a temperature approaching 0°C. The average depth of the deeper cores used in the SPECMAP timescale was 3230 m, where the temperatures are about 2.5°C (seawater freezes at about -1.7°C). This is not far above the depth occupied by Antarctic Bottom Water today. If only unmixed bottom water from the Antarctic Ocean occupied the depths of the world oceans, the water at the SPECMAP sites would have been about 3°C colder than today, causing an 0.8‰ positive bias in the $\delta^{18}O$ records. The compelling conclusion is that there was no warmer deepwater formed anywhere in the North Atlantic during the anomalous deglaciation. Antarctic Bottom Water then filled the world's oceans. Its low temperature during the deglaciation and the usual slight bioturbation that mixes the sediments masked the negative effect of the deglacial meltwater on the $\delta^{18}O$ values.

The deglaciation during an insolation minimum is a severe blow to the conventional Milankovitch wisdom, and was followed chronologically by another that was equally severe. Consider the low sealevel between the first and second high seastand (Fig. 16). For a long time even the existence of this drop in sealevel was doubted by many because it violates Milankovitch. Now, however, Esat et al. (1999) have shown that the world sealevel fell rapidly more than 80 m below the +7.4 m level of 136,700 yr BP. In 1999 they reported a collection of well-preserved shallow-water corals from Aladdin's Cave, a sea cave on the Huon Peninsula in New Guinea where the coral fossils were protected from tropical rains. Sea caves often form at times of nearly constant sealevel when underground freshwater, moving slowly through cracks in the limestone, exits into the sea not far from the shoreline. Seawater mixes into the freshwater in the passages near the exit point and the mixture is quite effective in dissolving the limestone in the passages because the mixture is undersaturated with carbonate. Aladdin's Cave therefore defines a world sealevel of close to -80 m that was probably a world sealevel minimum of that time. The four highest quality fossil corals from the cave were dated to a tight average of 130,400 yr BP. Elsewhere on the peninsula similar last interglacial deposits were found that had formed at the same elevation below present world sealevel.

Here we have reliable evidence that, after the anomalous high seastand at 136,700 yr BP, the volume of glacial ice increased rapidly and, in 7000 years lowered world sealevel by almost 90 m, or about the height of a 20 story office building. While this glacial growth occurred, the northern summer insolation was also rising, and at 130,000 yr BP (Fig. 8, at *a*) it had become higher than at the insolation maximum at 10,000 yr BP during the last deglaciation ! Again the natural world provided a direct and striking contradiction to the Milankovitch hypothesis. After 130,000 yr BP however, a rapid deglaciation began as insolation peaked at 128,000 yr BP and sealevel rose to a height of +5 m about 125,000 yr BP. One might cite this correlation of a rapid deglaciation with an insolation maximum in support of the conventional Milankovitch hypothesis, as has often been done in the past. But a correlation without a good causal connection can be misleading, and in Chapter 13, the case will be argued for a deglaciation in which Milankovitch insolation did not play a significant role.

The anomalous deglaciation during a period of weak insolation, followed by a reglaciation under strong insolation, would seem incredible to those committed to the Milankovitch concept, but the evidence cannot be denied. Here is probably what happened. The peak sealevel of +7.4 m requires the melting of an unknown amount of Antarctic ice, because even if all Greenland ice had melted (it didn't), world sealevel could not have gone as high as +7.4 m without a melting contribution from the Antarctic. This loss of Antarctic ice may seem odd, but in 1999 Hearty et al. reported evidence from the wind-blown dunes of carbonate sand on the Bahamas that an older sealevel of +20 m occurred between 400,000 and 500,000 yr BP, implying an even larger Antarctic deglaciation. The reason for the Antarctic ice loss will be discussed in Chapter 11.

The rapid increase in ice volume after 136,700 yr BP would be reasonable if the anomalous deglaciation had removed only, say, half of the large Laurentide Ice Sheet, with the remainder of the sealevel rise coming from loss of ice in Antarctica and Eurasia. Then, when climate conditions again favored glacial growth, the reglaciation could have been rapid due to the large residual nucleus of Laurentide ice. There is a more recent example of this. A similar rapid increase is well known to have occurred between 30,000 and 20,000 yr BP when sealevel fell from about -50 m to about -120 m, again suggesting favorable conditions for an ice-volume increase beginning with large nuclear ice sheets in North America and Eurasia. These conditions included a generous moisture supply inferred from the relatively warmer sea surfaces indicated by warm-water fauna in the northern gyre as discussed by Ruddiman and McIntyre in 1979.

With these contradictions in mind, it is clear that one or more previously unrecognized internal factors within the earth's climate system completely overpowered the strongest insolation variation, and the finger of suspicion points to the Mediterranean Sea. But we need to lay a good foundation for a discussion of the Mediterranean role in the anomalous deglaciation. First, a quick look at the dynamics of storms that supply moisture to the ice-sheet regions. Then a short discussion of conveyor-belt oceanography is appropriate, for there was a major change in ocean circulation and a reduction in the moisture supply to the ice sheets at that time.

It is also timely to indicate the unconventional direction that the arguments of this book are taking. Historically, the conventional wisdom has been that warmth and higher solar energy inputs determined the fate of the massive ice sheets, and that lack of warmth was responsible for their nucleation and growth. As the remainder of the book unfolds, it will become clear that it is, instead, the moisture supply to the ice sheets that governs their fate. Not only is the moisture needed in the form of snowfall to maintain the ice sheet and to enable its growth, but abundant atmospheric moisture brings heavy cloud cover, which can protect the ice from summer melting. Cloud cover in summer tends to occur where there are large areas of contrasting temperatures, such as the southern edge of the ice sheet at 0°C and a much warmer land surface to the south. Therefore, when the ice sheet is shrouded in clouds, melting is minimized, and the ice sheet tends to grow and advance. When clouds are absent, melting and deglaciation are accelerated. Sometime in the last million years, the North American ice sheet reached south as far as the outskirts of St. Louis, Missouri (Fig. 1), probably under a summer cover of clouds. And a good moisture supply from the Gulf of Mexico was not far away.

The actual amount of solar energy reaching the surfaces of the land, ocean, and ice sheets can be more strongly influenced by large differences in cloud cover than by differences in Milankovitch insolation. Just to toss a simplistic number out, a nearly continuous cloud cover through the summer months over a large region could reduce the incident ground-level solar radiation by over 60% by the reflection of sunlight off the cloud tops back into space. Compare this with a typical extreme variation of insolation at the top of the atmosphere of perhaps 15% at midsummer on June 21st due to orbital effects. Without a lot of moisture, there are few clouds. And the moisture is carried to the ice sheets from the ocean by storms.

8

Where the storms go

When determining the fate of the world's great ice sheets, there is nothing more important than getting the moisture to them to make them grow, or taking away the moisture supply to make them shrink. To get a better idea of how this process works, it is necessary to consider how the great masses of moving air are affected by the earth's rotation and how storms are generated. Storms are often steered in their paths by atmospheric jet streams, the high-altitude air flows that move approximately from west to east along zones of temperature contrast. Jet streams are a consequence of the earth's rotation. The rotation tends to change the flow direction of moving air or water anywhere on the earth except precisely at the equator. This is known as the coriolis effect, after Gaspard Coriolis, the French mathematician who first analyzed the consequences of the rotation on moving objects or fluids. One can understand the effect by an example, even without the mathematics.

Imagine you are in a plane traveling north over New York City at 500 km per hour heading for Albany, 230 km to the north. The New York point on the earth's surface is rotating eastward at a speed of 1050 km per hour. Very soon you would like to be over Albany, but there the rotational speed is only 1015 km per hour because Albany is closer to the North Pole and a little closer to the earth's polar axis where the speed of rotation is zero. If the plane's pilot did not correct the course, the higher eastward rotational speed carried from New York would have put the plane off course, and you would finish the flight to the east of Albany. The earth's rotation twists the northward movement to the right of whatever great circle route you are trying to take in the Northern Hemisphere, twisting to the east if traveling north and twisting to the west if traveling south. The rule is reversed in the Southern Hemisphere. The faster the ground speed is, the greater the tendency to be diverted from the desired heading. This effect increases with the sine of the latitude, trigonometrically, and is zero at the equator because a slight move to the north hardly changes the distance from the polar axis.

The coriolis rotational effect explains why Northern Hemisphere trade winds blow as they do. As air heated over the equator rises, the slightly cooler air at the surface to the north flows south to replace it, and that flow twists to the right. The result is a surface flow from northeast to southwest in the trade wind zone. These winds flow most strongly between 10° and 25° N latitude. The air that rises close to the equator moves back to the north and tends to sink in the 20°N to 35°N zone to complete the flow path. This vertical loop is known as the equatorial Hadley cell, and the zone of sinking air is sometimes called the horse latitudes (Fig. 17). The sinking air has had its moisture condensed out when rising over the tropics. Therefore, it is quite dry and this zone contains most of the deserts of the Northern Hemisphere.

In the trade wind zone of the subtropical North Atlantic, large-scale eddies are often amplified by the physics of water condensation. An eddy of air at lower altitudes that rotates counter-clockwise because of the coriolis effect will have a lower air pressure at its center

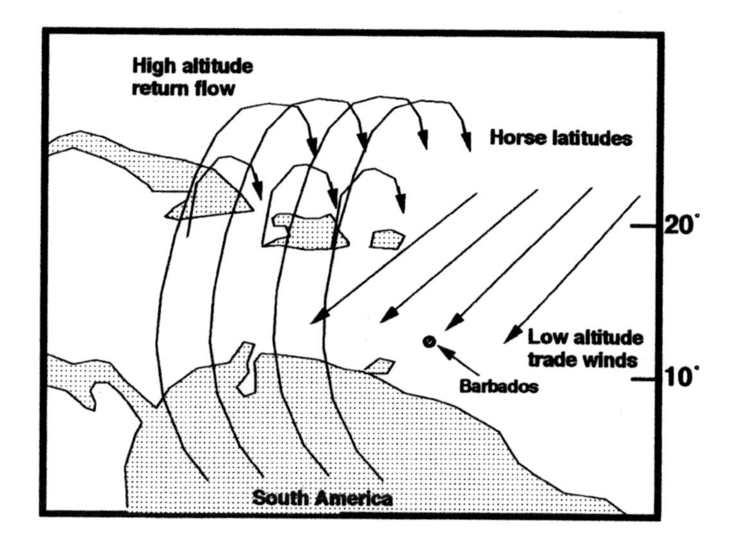

Figure 17: Atmospheric flow in the Hadley cell over the Caribbean. Heated tropical air rises and is replaced by surface flow of the trade winds. The high altitude return flow sinks over the subtropical latitudes causing dry climates there.

because the coriolis effect tends to move air away from the center. Over the warm ocean the humidity is high and the lower pressure within the eddy cools and condenses water vapor into cloud droplets in the lower altitudes of the central region. Energy given up by each water vapor molecule as it condenses on a droplet appears as heat (formally called the heat of vaporization), and consequently the air in the cloud is warmer than it would otherwise be, and becomes less dense and more buoyant than air outside the eddy. Therefore, the central air rises, condensing still more moisture. The rising air is replaced by air spiraling inward toward the center at lowest elevations near the surface. To conserve angular momentum, as the physics professors say, the rotational speed increases. This further lowers the pressure, condensing more water droplets and further amplifying the whole process. When carried to extreme, this is how hurricanes are born. They always require the coriolis effect and a moisture supply associated with a warm ocean. Consequently, they do not occur near the equator where the coriolis effect is small, and tend to occur less often in the horse latitudes where winds are calm and the air is dry.

The storms of the subtropics are interesting in themselves. Computer models show that they can also be highly effective in mixing very moist air northward into the midlatitudes and from there on into the high latitudes, thus supplying snow on the ice sheets. In 1989 Johnsen, Dansgaard, and White showed that values of excess deuterium (the heavier isotope of hydrogen) in snow on the higher elevations of the Greenland ice sheet could only be compatible with a high sea-surface temperature at the moisture source. The source is the warm Caribbean Sea, and the moisture is carried to Greenland by the same jet stream action that moves hurricanes northward along the east coast of the United States and Canada.

In the midlatitudes large-scale air movements are dominated by high altitude jet streams flowing above narrow zones of contrasting surface temperatures. Jet streams form as illustrated in Figure 18. The air over the warmer surface is heated and expands vertically, forming a column of air that is taller than a similar column nearby to the north above the colder surface. Consequently, the total

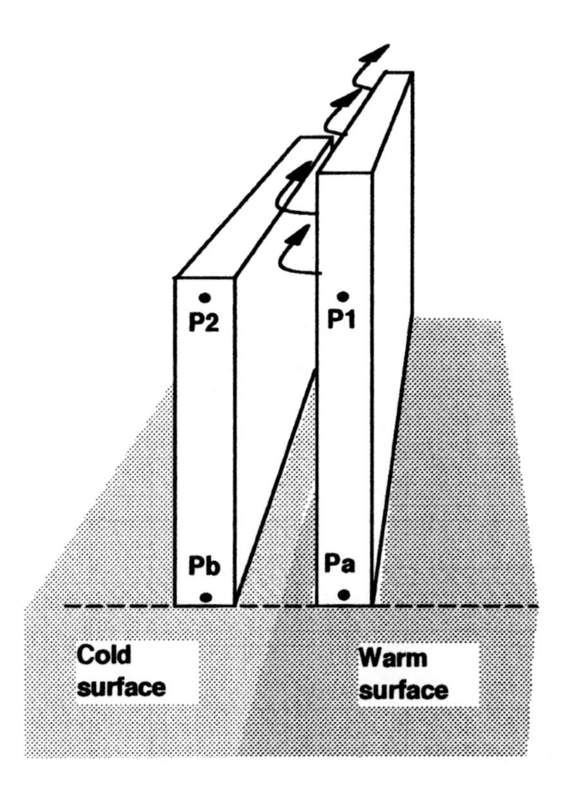

Figure 18: How atmospheric jet streams form in the Northern Hemisphere. Air over the warm surface expands, pushing the column of air upward. Consequently, air pressure at high altitudes in the warm column is greater than in the cold column (P1 greater than P2). The resulting pressure difference moves air toward the cold column, and that air flow is twisted by the coriolis effect to flow nearly parallel to the pressure gradient. Pa is slightly less than Pb at the land surface because a little air is transferred to the cold column.

mass of air lying above any given level at altitudes very much above ground level is greater in the warm than in the cold column. This causes a horizontal pressure difference that becomes quite significant at higher altitudes, and between 25,000 and 40,000 feet the pressure gradient tends to force air from the warm column over toward the cold one. But the coriolis effect comes into play, and much of the flow becomes twisted to a movement all along the zone of temperature contrast, resulting in a high-speed jet stream flowing generally eastward and almost, but not quite, perpendicular to the pressure gradient.

It is a fact that the paths of traveling storm systems are found beneath the jet stream, as can be observed by following modern weather maps over the midlatitudes of the Northern Hemisphere. But do the storms follow the jet stream or does the jet stream follow the temperature contrasts associated with the storms ? Perhaps a bit of both. Weather maps often suggest that a jet stream follows the contrasts along the storm front, yet jet streams flow even in the absence of a storm condition if there is a sufficiently strong temperature gradient over the earth's surface to drive the effect. One of the most persistent examples of a jet stream flowing along a boundary between contrasting temperatures is shown by the storm track path that runs northeastward over New England, northeastern Canada, Greenland, and Iceland and into the Nordic Sea. On the south side of the path is the warm Gulf Stream water, which flows from Florida northward along the coast to Newfoundland where it is deflected eastward. Farther to the northeast off Greenland the warm Irminger current keeps the sea-surface temperatures higher from Iceland to Cape Farewell at the southern tip of Greenland. The cold contrast to the north of the storm track is supplied by the cold land surfaces, the cold Canadian Current from the Arctic that is held against the coast by the coriolis effect, and finally the Greenland ice cap and the icy East Greenland Current along the Greenland east coast. Figure 19 from an early 20th century record by Klein shows some statistics illustrating the concentration of storms moving

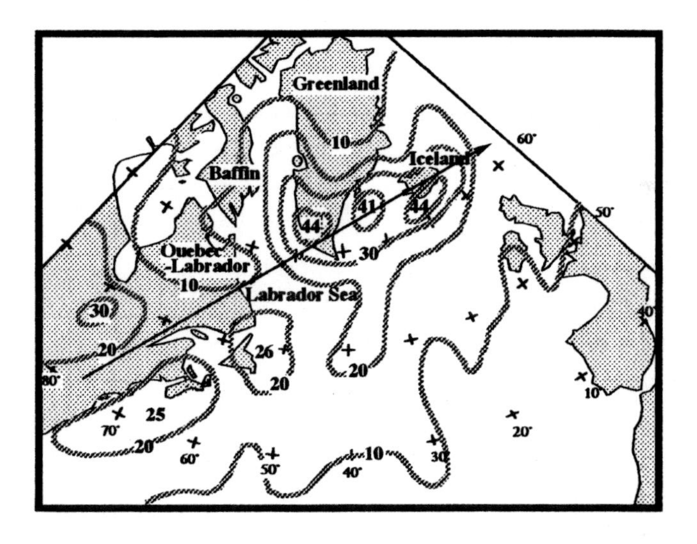

Figure 19: Some January statistics of traveling storms that follow paths along lines of strong warm-ocean/cold-land temperature contrasts. Arrow is the approximate path of the jet stream. Number of low pressure centers found in January over the years: 1909 to 1914 and 1924 to 1937. Modified from Klein (1957).

along this path. In the storm alley through New England, the 1917 meter-high Mt. Washington is well known for its frequent storms and high winds that occur throughout the year. The traveling storm systems follow this path more consistently than over a path on the continent to the west because the land-sea contrasts are large and more stable, whereas the temperature gradients over continental land are weaker and more variable. A similar land-ocean temperature contrast is associated with the warm Kuroshio current that flows northward east of Japan, and high-energy thunder storms moving along the front are common there.

The temperature contrast between the north edge of the cold Antarctic Water and the warmer mixed water to the north between Cape Horn at the southern tip of South America and the Antarctic continent causes strong storms to move along that oceanic front. In the days of sailing ships, such storms made "rounding the Horn" extremely difficult because of the westerly gales associated with the clockwise coriolis rotation of Southern Hemisphere storms. But in the Northern Hemisphere, the positions of oceanic fronts determine where the storms go that bring moisture to the ice sheets, causing them to grow or shrink. These fronts can be radically altered by changes in North Atlantic oceanic circulation.

9

A little conveyor-belt oceanography

There is a lot of chaos in the oceans. Describing ocean circulation is somewhat like predicting the weather. The general patterns of air circulation are known and can be predicted approximately, but when it comes to details, chaos reigns. Thus it is in the oceans. The great currents like the Gulf Stream that stand out clearly in our minds have their own unpredictable twists and turns, and the smaller currents are even more variable. Consequently, to grasp the main features of ocean circulation, measurements are averaged, sometimes over several years. It is therefore not surprising that oceanographers sometimes disagree as to the magnitude or placement of a flow in the ocean, and these uncertainties seem to contribute to the controversy on role of the Mediterranean Sea in the circulation of the North Atlantic. To consider this role, we need to start with a most fundamental oceanic factor: salt.

The North Atlantic is the saltiest of the world oceans. This is because it has net evaporation losses and is somewhat isolated from other oceans by North America on the west, Europe and Africa on the east, and shallow-water sills that partly separate it from the Arctic Ocean. The high salinity occurs because there are two avenues of water loss (Fig. 20). The largest loss is from the Caribbean, where water that is evaporated is carried by the trade winds westward over Central America into the Pacific, as discussed in 1968 by P. Weyl. The saltier water from the Caribbean circulates northward around the North Atlantic in the Gulf Stream. A somewhat smaller loss occurs when water in the Mediterranean Sea evaporates and is carried away eastward by the prevailing winds. The resulting Mediterranean water of higher salinity flows out through the Strait of Gibraltar and is distributed over the central and northern North Atlantic. In the winter the greater sea-surface salinity in the high-latitude North Atlantic that results from these two sources of excess salt allows denser surface water to sink when it is cooled, and the water that sinks, mainly in the Nordic Sea, is replaced by surface flow from the south. This drives the "conveyor-belt" oceanic flow, as W.S. Broecker (1999) of Columbia University calls it. This flow affects oceans throughout the world, and is particularly important in providing Europe with a mild climate, in contrast to colder northeastern Canada.

To the east, the climate in Great Britain, Scandinavia, and Europe is rather mild and winters are not severe because these regions are downstream in the air flowing over the warm parts of the conveyor-belt oceanic flow. Upstream to the west at similar latitudes, summers are cool and short and winters can be quite cold in Labrador and northeastern Canada. The warm part of the conveyor-belt is known as the North Atlantic Drift, and it originates as a piece of the warm Gulf Stream (Fig. 20). The northward flow of this Drift water splits into two branches. One branch, the Norwegian Current, flows northward west of Norway, and today its warmth keeps the Arctic seas relatively ice-free as far north as Spitsbergen in the Svalbard Archipelago, and eastward far beyond northern Finland.

The other, the Irminger Current, crosses the Atlantic south of Iceland, moving west and mixing as it goes. A bit

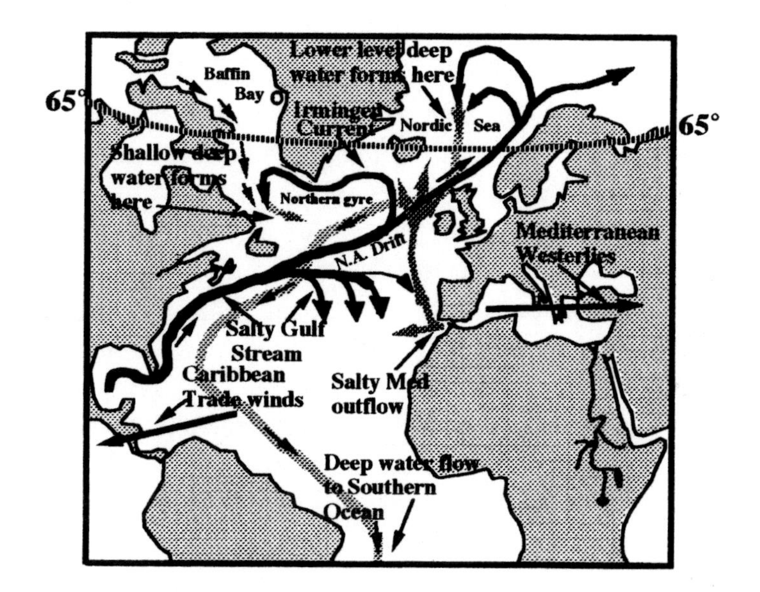

Figure 20: A salt-driven conveyor-belt oceanic flow. Water vapor is carried by persistent winds out of the Caribbean and the Mediterranean Seas. This loss increases the salinity of surface water, which moves northward. When cooled in winter it becomes dense and sinks far below the surface in the Labrador and Greenland-Norwegian Seas (the Nordic Sea). The sinking water keeps the world's deep oceans slightly warmer than they would otherwise be by competing with colder and less-saline deep water that sinks in the Antarctic.

of it reaches southern Greenland and continues northward as the West Greenland Current along the west coast of Greenland as far as Disco Island. Its regional warming effect is small and is opposed by the cold Labrador Current (sometimes called the Canadian Current), which consists of water from the Arctic Ocean that filters through the channels around Ellesmere and Devon Islands, and hugs the Baffin and Labrador coasts southward. This cold current tends to make the climate cold along the Labrador coast and in the remainder of northeastern Canada. Much of the West Greenland Current mixes with the Canadian Current, returns southward, and the combined flow turns eastward north of the Gulf Stream, joining the North Atlantic Drift to complete a large circular flow called the northern gyre. By noting small differences in surface salinity in the 1970s Dickson et al. (1988) traced a lower-salinity mass of water around the gyre in about seven years, which is an indicator of the time needed for possible abrupt circulation changes with climatic effects.

The gyre is, of course, not a closed circulation pattern. Outside water enters and inside water leaves. A very significant part that leaves consists of water that sinks to deep levels in the Nordic Sea and shallower levels in the Labrador Sea when surface water is cooled enough in winter to make it denser than the average of water in neighboring areas. Only a small density difference is needed for surface water to sink and this difference is enabled by the higher salinity of the source water, at least half of it originally from the warm and saline Gulf Stream. Salt molecules are more than three times heavier than water molecules, so the greater salinity brings about a greater density when winter water temperatures are the same. The water that sinks from the surface is replaced by saline surface water moving northward on the east side of the Atlantic from the lower latitudes. The process of sinking and replacement that drives the conveyor-belt is a simple approximation for what is actually a quite complex oceanic circulation system. The conveyor-belt is responsible for the European warmth because warm and salty subtropical Gulf Stream water is continually injected into the eastern part of the gyre to replace water that reaches the Nordic Sea and sinks into the deeper ocean. Without this deepwater formation, Europe would be quite cold.

The surface water entering the Nordic Sea is more saline than water entering the Labrador Sea, which has mixed to a greater extent with somewhat deeper water. When winter-cooled water becomes dense enough to sink off Labrador, it goes down hundreds of meters and becomes part of the intermediate-level North Atlantic. Nordic Sea water has more salt in it. Its temperature is somewhat higher at the time of sinking, and when fully cooled its density is greater. Therefore, it can continue to sink as it mixes with surrounding water after pouring over the Faeroe-Scotland sill, and can fall to the 3000 m-depth. It moves on into the Western Boundary Current, circulates across the equator to the Southern Ocean (Fig. 20), and from there goes widely throughout the world, keeping deeper levels of the oceans warmer than they would otherwise be. The Labrador Sea deepwater likewise contributes to the flow to the Southern Ocean, as discussed by Lehman and Keigwin in 1992.

Surface water around the Antarctic continent also cools and sinks in the frigid winter climate and, because its salinity is quite a bit less than in the North Atlantic, the Antarctic water going to the bottom is much colder. Therefore, without the somewhat warmer and more saline deepwater formed in the North Atlantic, most of the world ocean would be a few degrees centigrade colder than today. These two sources of deepwater compete with each other and this may lead to minor widespread climate oscillations. This effect is called "the bi-polar seesaw" and is the subject of much ongoing oceanographic research, as discussed by Seidov and colleagues in 2001.

The deep- and intermediate-level water in the Southern Ocean in the Atlantic sector is probably about 2°C warmer than it would be without the input of saline North Atlantic water. As described in Deacon's 1977 review, this mixed water from the north rises to the surface around Antarctica to replace cold water that leaves that region. As it rises it brings up dissolved nutrients that nourish enormous numbers of krill that feed the populations of seals, penguins and whales in Antarctic waters. This rising flow can be thought of as another conveyor-belt that splits into two cold branches at the surface (Fig. 21). One branch forms mainly in the winter when sea ice freezes. In the freezing process, sea salt is not incorporated into the ice, and the denser and slightly more saline water mixture that remains sinks to the deepest levels close to the continent and forms Antarctic Bottom Water. This water mass flows into all of the deep basins in the world ocean. The other branch originates in the summer when the Antarctic sea ice melts. The resulting less dense and lower-salinity meltwater mixture at the surface is the Antarctic Water that accumulates in a broad area around the continent. The greater density of the more-saline water beneath the Antarctic Water reduces thermal convection of warmer water from below and makes possible the surface cooling and freezing of sea ice during each following winter season.

The winter storm tracks in the Southern Ocean tend to circle the continent in the zone of contrasting temperature between the frigid continental ice sheet and the edge of the Antarctic Water at the Antarctic Convergence Zone, where

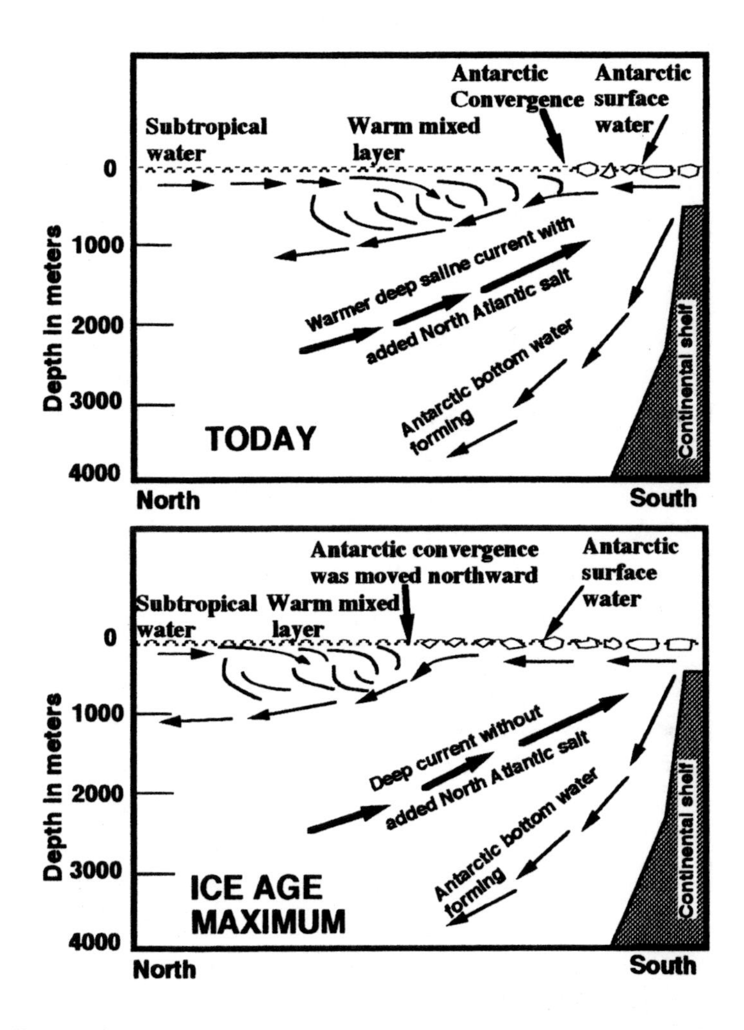

Figure 21: Vertical circulation in the Southern Ocean. The Antarctic Convergence marks the northern limit of heavy sea ice. When less deepwater is formed in the North Atlantic, the cold Antarctic surface current becomes less salty and not as dense. This pushes the Antarctic Convergence far to the north, making Southern Hemisphere climate colder. See text.

the warm mixed water to the north begins (Fig. 21). Consequently, over much of the zone of Antarctic Water the strong storm winds blow from the west, while closer to the continent weaker winds blow from the east. This broad band of westerly winds drives the surface water eastward, causing a dominant circumpolar flow around Antarctica.

Because the cold Antarctic Water is slightly more dense than the mixed subtropic water to the north, there is a pressure gradient at the border between the two types of water that causes the heavier Antarctic Water to slide to the north beneath the warmer and less dense mixture at the convergence line. The transition from cold Antarctic Water to warm mixed water is quite abrupt as seen at the surface. The air above the Antarctic Water can be quite cold and wintry, while a few kilometers to the north of the convergence the air is mild and pleasant, "like spring in England" as Deacon puts it. As the more-dense Antarctic water slides beneath the surface, it mixes with the warmer subtropical water, thus cooling it and increasing its density as it approaches the convergence line. The density difference is therefore rather small at the convergence line, and this leads to an explanation for the large northward extension of heavy sea ice that was inferred from cores in the Atlantic and Indian Ocean sector of the Southern Ocean near the last glacial maximum, when little northern deepwater was formed.

About 18,000 yr BP, near the approximate glacial maximum, a large release of icebergs from northeastern Canada made sea-surface temperatures across the midlatitude North Atlantic very cold. There was no North Atlantic Drift. This implies little or no North Atlantic

Deep Water formation. With no high-salinity deepwater from the North flowing into the Southern Ocean, the density of the Antarctic surface water at today's latitude of convergence would have become less than otherwise. Consequently, it would not have been able to sink below the subtropical water at that latitude. Instead, the layer of Antarctic Water accumulated to a broader extent in the surface zone around the continent. This would have pushed the convergence line northward to a warmer latitude where less sea ice freezes. The resulting higher salinity and density of the Antarctic Water there would have again allowed it to sink below the subtropical water (Fig. 21). This mechanism explains the result of studies of abundances of diatoms in sediments by Hays et al. (1976b), which show that, near the last glacial maximum, permanent summer sea ice in the cold Antarctic Water extended about 10° of latitude (1110 km) north of today's melt-back limit in the South Atlantic and Indian Oceans.

The importance of this mechanism should be emphasized, because it implies that variations in North Atlantic Deep Water formation have a direct and large effect on the climate in the Southern Hemisphere. The areas that are at least partly covered by sea ice in winter today extend outward from the Antarctic continent by as much as 1500 km. Most of the ice melts back in summer. During the interglacial climate of today, a large fraction of deep-sea sediment off Antarctica consists of microscopic diatoms. Diatoms are plants that require the sunlight in the surface water. Hays and his colleagues showed that during the last ice age, and particularly near the maximum about 18,000 years ago, diatoms were scarce, and the sediment consisted mainly of clay. Sunlight could not penetrate through the sea ice and maintain a diatom population. This implies that the heavy sea ice then did not melt back during the summer.

The heavy sea ice reduced the transfer of heat from the ocean to the air, and the smaller amount of absorbed solar insolation due to the larger area of cloudy skies over the cold Antarctic Water would have had a significant climate cooling effect. The northward extension of the Antarctic Convergence Zone would therefore have cooled all the southern latitudes, and heavier sea ice would have moved storm tracks away from the continent. This reduced the heat transported to the continental ice sheet and cooled it by at least 6°C at altitudes where snow crystallized out of the atmosphere. The magnitude of these effects can dominate the Milankovitch variation of the southern insolation, and is the likely reason that glacial climates are synchronous in the Northern and Southern Hemispheres.

Returning now to the North Atlantic, Broecker's salty conveyor-belt is commonly depicted as the North Atlantic Drift off-shoot from the Gulf Stream. However, to address the perplexing contradictions of the Milankovitch hypothesis, an additional mechanism for northward salt transport is needed, and we must consider the Mediterranean Sea and its salty outflow at the Strait of Gibraltar. There is a deep current that flows out across the sill in the strait. It is about 1.7‰ more saline and substantially more dense than the Atlantic water above that is flowing into the Mediterranean (Fig. 22). This salinity difference is less at the sill than the 2.2‰ difference between the 38.4‰ outflow water approaching Gibraltar from the east and the inflow water from the west because

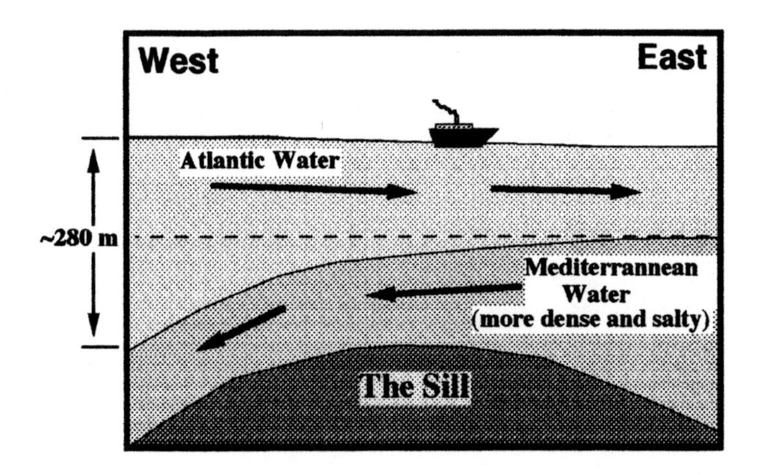

Figure 22: Exchange currents at the Gibraltar sill that occur because of the greater density of Mediterranean water caused by its higher salt content.

some counter-current mixing occurs across the interface between the two flows. The outflowing current velocity at the 280 m-deep sill in the strait is about 1.0 m per second, and the remainder of this chapter is devoted to the fate of this outflow and its load of excess salt.

The magnitude of the outflow is difficult to measure precisely, because it varies with tides and atmospheric pressure changes, but an estimate based on data from Bryden and Kinder's 1991 paper suggests an average outflow of about 790,000 cubic meters per second (m^3s^{-1}), assuming a steady-state salinity. For comparison, the average flow of the Nile River entering Lake Nasser above the Aswan High Dam over the last few decades is about 2800 m^3s^{-1} according to Gasser and El-Gamal (1994). The Mediterranean outflow passes out over the Gibraltar sill in the depth zone between 140 and 280 m, and because of its higher density it flows rapidly down the slope of the continental shelf toward the northwest, mixing with the much colder water of the deeper Atlantic and leaving a trail of mixed water at each depth as it sinks. Eventually most of it attains a buoyant equilibrium where the mixture lies in a depth zone between 800 and 1500 m at a temperature of 6-10°C in the lower thermocline, which is the depth range of steep vertical temperature change in the upper ocean. The mixture in this zone is identified by data from sediment cores on the continental slope off Portugal reported by Schönfeld and Zahn in 2000, and is clearly defined by its higher salinity measured in depth surveys of the open ocean, as depicted at different depths in Worthington's 1976 book: *On the North Atlantic Circulation.* The idealized directions of flow for the outflow mixture components as a function of buoyant equilibrium depth are illustrated in Figure 23.

J. Reid (1978, 1979) has published two papers that discuss the patterns of salinity in the upper levels of the North Atlantic, with emphasis on the contribution to the high-latitude North Atlantic sea-surface salinity made by the Mediterranean outflow. His controversial proposal is that its contribution of salt to the Nordic Sea was a strong factor in the formation of North Atlantic Deep Water. Some Mediterranean water is found along the African coast as far south as 24°N latitude, and is apparently carried there by the surface current flowing southward off Portugal. At slightly lower depths of 500-1000 m in the Reid data, a wedge of higher-salinity Mediterranean water mixture extends across the Atlantic west of Gibraltar. The

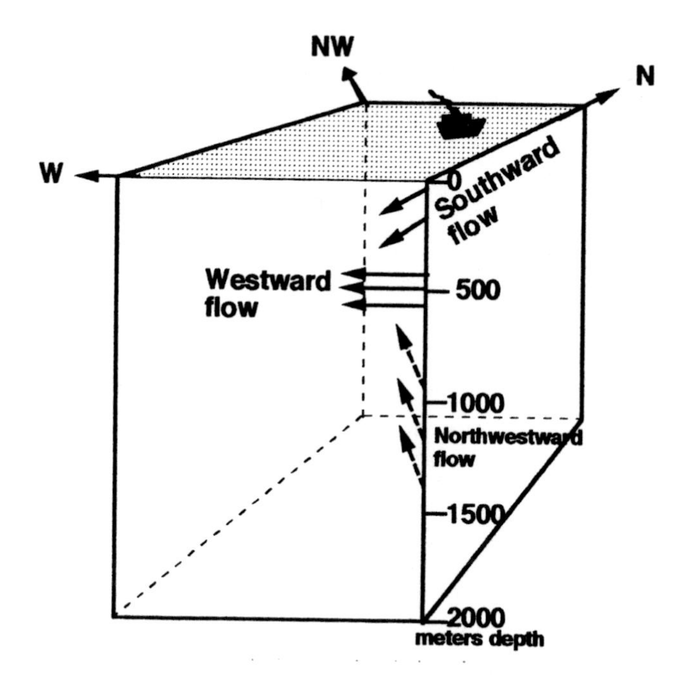

Figure 23: Mediterranean outflow directions versus depth of outflow mixture. The dense saline outflow mixes with Atlantic water as it sinks, shedding mixtures at all levels. See also Reid (1994), in which the deeper northward flow is shown in Figures 8f and 13b.

excess salinity attributed to the Mediterranean outflow in the wedge diminishes in the westward direction, partly because diffusion processes add outflow water to the North Atlantic Drift water lying above it. Nearer the Strait of Gibraltar, Pollard and Pu in 1985 showed the striking difference between the salinity maximum of this mixture at about 900 m-depth and a separate upper salinity maximum at the surface due to the overlying Gulf Stream water. The deeper salinity maximum extends northward in a wide band.

Reid says this band represents a flow that rises to shallower depths due to the geostrophic effects (the coriolis effect of rotational twist), where it mixes into the surface levels west of Scotland. There some it flows across the Faeroe-Scotland sill into the Nordic Sea. The flow is not a broad steady current, however. Otto and Aken in 1996, using modern radio-equipped drifters, reported very erratic patterns of flow in the area south of Iceland and the sill (Fig. 24). The mixing that is implied by this flow pattern is not limited to the surface layers, and is found also to the south as the Mediterranean water mixture rises beneath the North Atlantic Drift, as shown in Figure 9 of Reid's 1979 paper. As the water moves northward between latitudes 40°N and 55°N, the deeper salinity maximum of Mediterranean water rises and merges with the surface salinity maximum of the North Atlantic Drift water. This suggests a substantial contribution of Mediterranean water in the flow to the Labrador Sea as well as to the Nordic Sea.

Some oceanographers dispute the importance of the Mediterranean water in this region. Nevertheless, Reid has a strong case for a good flow of Mediterranean water into the Norwegian Sea, based on measurements of trace amounts of silica in seawater at various locations in the North Atlantic. Like many other chemical compounds and elements, silica is held in solution in small amounts in the world oceans. The silica is used by diatoms, microscopic plants that live at the ocean surface by photosynthesis. They extract silica from the surface water to incorporate it into their skeletons, and diatoms are plentiful in the mineral-rich water upwelling around Antarctica (Hays et al., 1976b). Consequently, bottom water in the Antarctic is quite rich in dissolved silica because small amounts dissolve out of the abundant sedimentary skeletons of dead diatoms that lived at the surface. The deepest water in the North

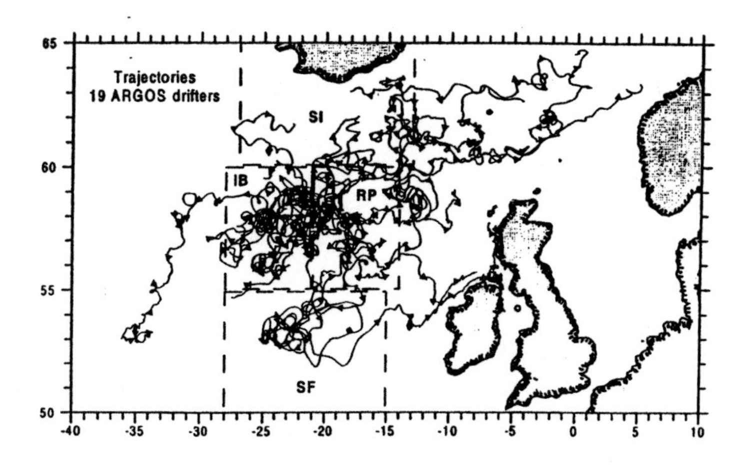

Figure 24: Chaotic flow south of Iceland shown by 19 drifters equipped with global position sensors. Average flow is to the northeast. This is an area where upwelling Mediterranean water identified by Reid would be mixing with North Atlantic Drift water. From Otto and van Aken (1996), used with permission of Elsevier Science.

Atlantic below 4000 m-depth is Antarctic Bottom Water, with its heavy load of trace silica. Diffusion processes mix the silica upward to the surface in the North Atlantic, but the dissolved silica concentration in the near-surface water is quite low nearly everywhere because the diatoms living in the surface water continually remove it to incorporate it into their skeletons.

In contrast to most locations, Reid found the trace silica concentration to be high near the Faeroe-Scotland sill, where rapidly falling sea-surface temperatures as the sill is approached from the south suggest upwelling of cold water from below the thermocline, hundreds of meters below the surface. In the sea surface over the sill, the silica concentration was about four times greater than in more southerly waters of the North Atlantic Drift and other surface areas of the Atlantic. High concentrations were also found nearby across the sill farther to the north in the Norwegian Sea Current. The straightforward interpretation is that silica from the Antarctic Bottom Water diffused up into the Mediterranean mixture at the 800-1500 m-depth, was carried northward, and upwelled to the surface near the sill at a rate too fast for the diatoms to remove it. This is convincing evidence that there is a mixture of Drift water and Mediterranean water passing over the sill, and a substantial part of the water entering the Norwegian Sea between the Faeroes and Scotland appears to be the Mediterranean water mixture. Reid therefore had good reason for his proposal that, without salt from the Mediterranean outflow reaching the Nordic Sea, the warming of the European climate by the conveyor-belt circulation would be much less.

His argument is strengthened by the work of Greatbatch and Xu, published in 1993. Their results make it possible to estimate approximately the proportions of excess salt carried by the North Atlantic Drift and by the Mediterranean mixture into the higher latitudes north of Ireland. From a network of densities derived from salinity and temperature measurements over the high latitude North Atlantic, they plotted, for two time intervals, average northward flow velocities as a function of depth across a vertical plane in the ocean westward from Ireland at 54.5°N latitude. A redrawn plot of average flow for the years 1970-1974 is shown in Figure 25. There are clearly two different flow regimes crossing this vertical plane, each having a velocity maximum. The flow on the west is the

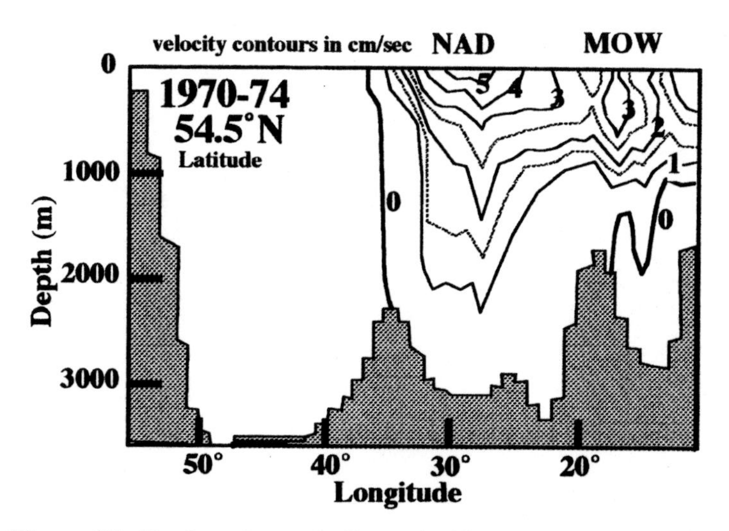

Figure 25: Northward oceanic flow velocities at 54.5°N in a vertical cross section west of Ireland. The North Atlantic Drift appears to be largely a separate flow relative to the Mediterranean outflow mixture. Modified from Greatbatch and Xu (1993) with permission of The American Geophysical Union.

main North Atlantic Drift current and it has highest velocities at the surface. The northward flow on the east occupies the band of higher-salinity Mediterranean water identified by Reid. Its maximum velocity is found below the 100-200 m-thick mixed surface layer and corresponds to the core of maximum salinity in the Mediterranean mixture. Much of the surface water above this core may indeed be part of the North Atlantic Drift, but most of the water below about 300 m is probably Mediterranean water mixing into the Drift water above it and moving northward at a slightly higher velocity than the adjacent non-Mediterranean water, probably because when the outflow sinks to the 800-1500 m-depth west of Gibraltar, it imparts momentum to the water with which it mixes.

If the Greatbatch and Xu velocity field is a reasonable picture of the northward flows of the North Atlantic Drift water and the Mediterranean mixture, it is possible to make an estimate of the proportions of excess salt carried by the Drift and Mediterranean mixtures as they move across latitude 54.5°N. The salinity as a function of depth can be roughly approximated from Figures 23, 25, and 27 in Worthington's 1976 book. In Worthington's figures the topmost mixed layer at the surface of the ocean has the higher salinity of Gulf Stream water at the 54.5°N latitude of the cross section of Greatbatch and Xu. Below the mixed layer from 200 m-depth on down, salinities are sharply lower and it seems that the Mediterranean outflow mixture at these depths dominates over a broad area of the ocean to the west of Ireland. By summing the product of velocity and excess salinity for small incremental areas in Figure 25, the relative rates of excess salt carried north in the upper levels can be estimated. During the 1955-1959 interval the proportion of the excess salt moving north that can be attributed to the Mediterranean outflow was about 42%. During 1970-1974 it was about 37%. These estimates do not take into account any saline outflow water from Reid's intermediate-level wedge that may contribute to the North Atlantic Drift.

During glacial times, less deepwater formed in the North Atlantic and the CLIMAP studies suggest that the North Atlantic Drift was weak, although the excess salt in the Mediterranean outflow may have continued without major change. Therefore, the Mediterranean contribution of salt was probably mainly responsible for the smaller

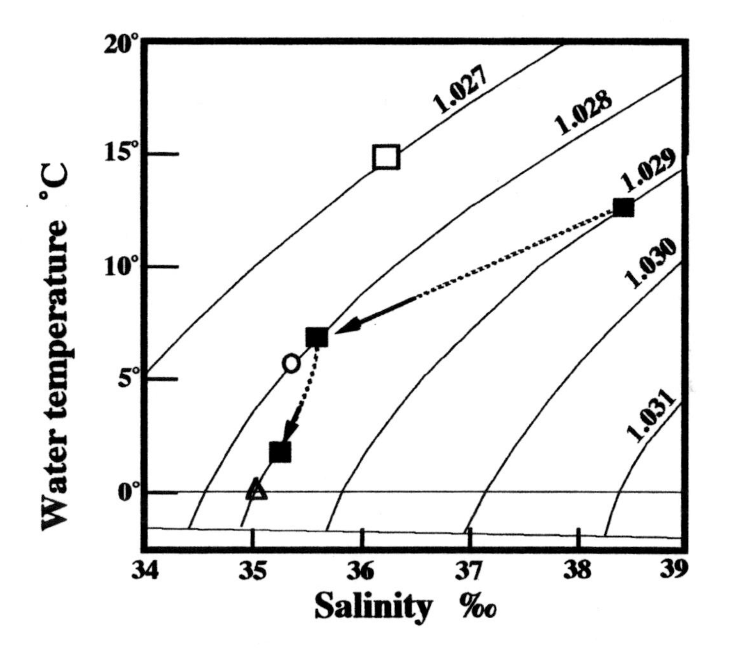

Figure 26: Modification of Mediterranean outflow properties of temperature, salinity, and density as it mixes and is exported to the Southern Ocean. See text. Dark squares - Med outflow and subsequent mixture. Open square - surface water west of Gibraltar into which the outflow begins to mix. Open circle - North Atlantic water at equilibration depth before mixing with outflow. Triangle - cold Antarctic Water overlying upwelling mixture around Antarctica. Density curves, in grams per cubic centimeter, from American Institute of Physics Handbook (1957).

amount of deepwater that formed in normal glacial times. It is instructive to consider in Figure 26 how the Mediterranean outflow mixes and has its salinity, temperature, and density modified during glacial times when it is the main source of excess salt exported to the seas around Antarctica. Neglecting the slightly greater salinity of the world ocean then, the water starts at 1.029 density and at a temperature of 12.6°C. The outflow sinks rapidly and equilibrates with water in the 1000-1500 m-depth zone at a typical temperature of 6°C. On a much slower timescale, the outflow subsequently mixes still more as it rises up toward the surface in northern high latitudes. There it is winter-cooled and sinks to intermediate levels, and after further modification, reaches the Antarctic seas. By that time it is only slightly more saline than the surface water that it upwells to replace. After upwelling, it forms new cold Antarctic Water and Antarctic Bottom Water as described earlier in this chapter.

In Chapter 11 we will see that, during the anomalous deglaciation that began at 144,000 yr BP, the exchange currents at Gibraltar were reversed. Large amounts of glacial meltwater mixed with salty water from the Black Sea and poured into the Mediterranean, and only low-salinity water entered the Atlantic at Gibraltar. There was no Mediterranean contribution of saltier water to the high-latitude North Atlantic, and the alterations in climate that resulted were world-wide. Before proposing an explanation for this remarkable change, we will consider the earlier history of the Mediterranean Sea from its birth to its death, and its rebirth and development into the body of water that we know today.

10

Why the Mediterranean is what it is and is not what it was

The Mediterranean is the world's largest body of water having a salinity greater than the North Atlantic. It has an area of 969,000 km^2, and connects to the world ocean only through the Strait of Gibraltar (Fig. 27). It does, however, connect to the Black Sea through the narrow Bosporus Strait, which on rare occasions is itself a connection of some climatic importance. The Mediterranean is now only

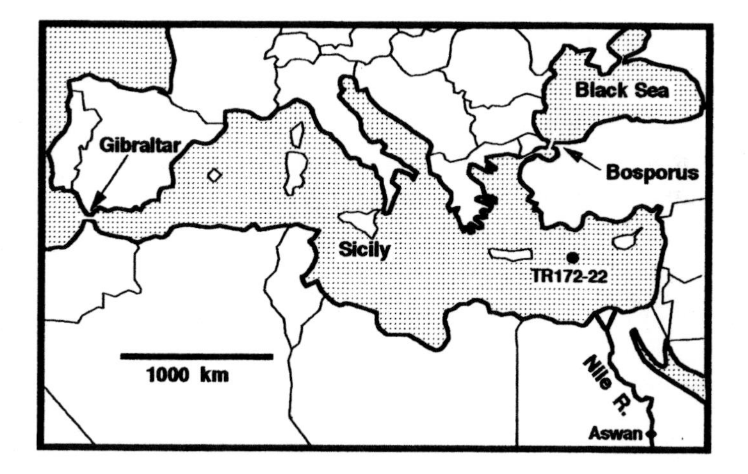

Figure 27: The Mediterranean and Black Seas with connections at the Bosporus and Gibraltar Straits. Core TR172-22 provided a record of Eurasian meltwater effects when meltwater flooded the Mediterranean and shut down North Atlantic conveyor-belt circulation 144,000 years ago.

Post press correction: The area of the Mediterranan Sea is 2,510,000 km^2

a remnant of what was once the great Tethys seaway separating the continents of Africa and Eurasia. The Tethys Sea was strangled by continental drift, and no discussion of climate should fail to acknowledge the role of continental drift. This was another of those ideas that fell outside the conventional wisdom of its time. It was forcefully advocated by Alfred Wegener (1966 translation), an engineer who in the 1910-1920 period became intensely interested in the natural history of continents. He suggested that the match in the shape of the coastlines of North and South America with the coasts of Europe and Africa occurred because they were once connected and have since drifted apart. He published this proposal, which was strengthened by describing the similarity of plant and animal types in America and Europe, a fact that had been conventionally explained by the hypothesis of ancient connecting land bridges that had since sunk beneath the sea.

Wegener failed to convince his contemporaries because he was unable to suggest a mechanism to drive the continental movement. The drifting continent idea was also deemed impossible by the opponents who pointed out that the viscosity of Earth's crust was too great to enable drifting to occur. Now we know more than they knew in Wegener's time. The viscosity of the hotter deeper rocks below the crust is much less than viscosity of the thin crust, and deep thermal convection currents circulate hot rock from down in the mantle up to the crust, over, and down again, carrying the continents around on the surface like scum on water that is spot-heated in a saucepan - on a timescale of hundreds of millions of years. Although his arguments were substantial, his drift concept was scorned for over thirty years until evidence for seafloor spreading was found in measurements of Earth's past magnetic field reversals in the frozen lava rocks at the bottom of the sea.

As the hotter and less dense rock in the earth's mantle rises toward the surface, it bulges up the seafloor to form the midocean ridges (Fig. 28). These giant convection currents spread the seafloor apart at the ridge. The lava that erupts in the great crack along the spreading line is soon cooled by seawater, and as it cools, the earth's magnetic field aligns the magnetic minerals with the field, leaving a frozen record of the direction of the magnetic field at that time. For reasons not clearly known, the magnetic

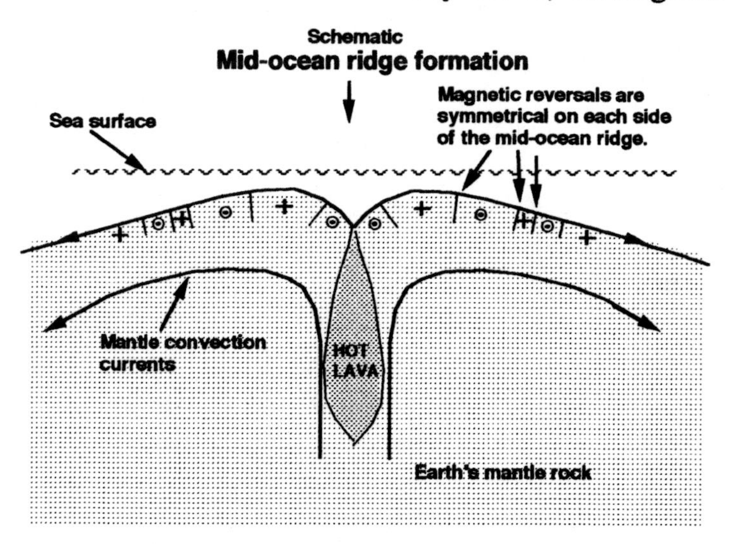

Figure 28: How midocean ridges form. Rising convection currents in Earth's mantle bulge up the seafloor. Magnetic field orientations of minerals (arrow tails and heads) in the frozen lavas reveal the spreading history of the ridge. The mirror-image record of reversals in the earth's magnetic field on each side of the ridge was a convincing factor in the acceptance of continental drift.

poles in the deep core of the earth occasionally reverse. Magnetic south becomes north and magnetic north becomes south. These reversals are recorded in the frozen lavas, which are slowly carried away from the central ridge, leaving a symmetrical record of the switches in the magnetism of the earth over millions of years. The record of these reversals can be read by a magnetometer on board a ship cruising across a midocean ridge having its axis oriented north to south. This type of evidence for continental drift was totally convincing.

Inexorably over millions of years, movement of the great continental plates moved Africa and the Near East toward a contact with Eurasia. The eastern end of the old Tethys seaway disappeared first, leaving the remnant pits of the 2000 m-deep Black Sea and the shallower Caspian and Aral Seas. As Africa approached Europe the crust buckled downward and created two other deep pits in the eastern and western Mediterranean, separated by a rather shallow sill around Sicily. Sedimentary rocks of southern Spain and Morocco show that, up to about 6 million years ago, the Mediterranean Sea of that time was connected to the Atlantic through the Betic Strait, a fairly broad channel north of Gibraltar, and the Riff Strait to the south of Gibraltar, the only channels to the outside ocean.

About 6.2 million years ago, crustal movement raised the Betic and the Riff Straits above sealevel and isolated the Mediterranean from the world ocean, a story told by Hsü et al. in 1987. This isolation lasted slightly more than a million years. Because of the dry Mediterranean climate, water lost by evaporation exceeds the inputs of rivers and rainfall, and the Mediterranean basin was converted into an enormous desert having only a few salty lakes with their surfaces far below sealevel. The layers of salt found below the sediments on the bottom of the Mediterranean today testify to the saturation of the lakes with salt that crystallized out and formed the salt beds under hot and arid conditions. Blanc and Duplessy in 1982 showed that the loss of the salty outflow from the Mediterranean during this time eliminated deepwater formation in the North Atlantic and caused the deeper water to flow from south to north. This is consistent with a single supply of deepwater formed in the Antarctic, and is a reversal of the direction of the flow occurring in all but the deepest depths when deepwater is forming in the North Atlantic. The Blanc and Duplessy result strongly supports Reid's contention that Mediterranean saline outflow is important for the formation of large amounts of North Atlantic Deep Water.

Almost five million years ago, further crustal movement lowered the land surface to just below sealevel south of Gibraltar, and Atlantic water poured into the western Mediterranean basin. No doubt seawater entered at a low and intermittent rate at first due to tidal action. As flow increased, the torrent carved out a deep valley that is now submerged as part of the seabed on the east side of Gibraltar. When the western Mediterranean became nearly full, it overflowed the sill at Sicily into the 2000 m-deep basin of the eastern Mediterranean. There the torrential inflow carved a narrow canyon about 800 m-deep between Sicily and Italy. For tens of thousands of years during the transitional times as the early Strait of Gibraltar was deepening, these spectacular saltwater cataracts would have been unique features of the Mediterranean scene.

The Strait of Gibraltar continued to deepen. The inflow increased enough to fully compensate for evaporation losses, and about 4.8 million years ago the Mediterranean sealevel became nearly equal to the Atlantic west of Gibraltar, and the exchange currents through the strait then became established (Fig. 22). Initially, with only a small shallow channel to the Atlantic to flush out the Mediterranean brine, the salinity of the outflow would have been very high. As the strait slowly deepened and widened, the increasing flow of less dense and less salty Atlantic water through the strait would have lowered Mediterranean salinity to that of today, although significant ice-age fluctuations of salinity continue to occur and are the theme of this book.

Strange as it seems, the total volume rate of river inflow to the Mediterranean is now only about 1% of the deep outflow at Gibraltar. The shallow inflow from the Atlantic is slightly greater than the outflow, and this compensates for Mediterranean evaporation losses. The difference between the larger amount of water lost by evaporation and the smaller amount the Mediterranean Sea receives in the form of rainfall and river inputs is called the hydrologic deficit. This net loss increases the salt concentration in Mediterranean water, and as noted earlier, the Western Mediterranean salinity at the sill is about 1.7‰ higher than the salinity of inflowing Atlantic water above it. This puts the denser, more saline Mediterranean water at the Strait of Gibraltar at the 280 m sill depth under greater pressure than water at the same depth a short distance away in the Atlantic. Consequently, the heavier Med water pours out over the sill. The water leaving through the strait lowers the sea-surface level in the Mediterranean slightly, creating a pressure gradient that drives Atlantic surface water eastward in the 0-140 m-depth to replace the outward flow and maintain a constant Mediterranean sealevel.

It will be useful to note that in the Bosporus Strait there are also exchange currents that carry dense eastern Mediterranean water of 39‰ salinity northward to the Black Sea along the bottom of the strait, and less dense Black Sea surface water having a salinity of about 17‰ southward to the Mediterranean. The invisible deep current was known to early mariners, however, who are said to have attached ropes to baskets of stones and lowered the baskets into the deep flow to pull their boats northward against the surface current.

In addition to the Mediterranean water balance, there is a salt balance, because, in a steady state with no increase or decrease in Med salinity, as much salt in the outflow must leave as enters in the inflow. So the slightly smaller volume of outflow carries a higher concentration of salt to balance the salt budget. But these two balancing acts occur on quite different timescales. The water balance merely requires movement of masses of water in or out through the Strait of Gibraltar and the change occurs quite rapidly. In the event of a sudden loss in river input, for example, the loss of the Nile River flow to the Mediterranean caused by the completion of the Aswan High Dam, the Mediterranean sealevel would have fallen to a new slightly lower sealevel, perhaps in a matter of weeks, with an increase in the inflow at Gibraltar to compensate for the Nile loss. On the other hand, with the loss of the Nile freshwater input, the salt balance requires that the salt concentration increase throughout the volume of the Mediterranean Sea. This is a much slower change because the buildup of salt to a new higher equilibrium salinity is slowed by the loss of salt in the Mediterranean outflow at Gibraltar. This resembles a textbook concentration and mixing problem. In the simplest model where the Mediterranean is one large basin and the exchange currents are constant, the asymptotic time needed to reach 63% of the final salinity is about 120 years. However, the change is more complicated than in the model because as the salinity increases, the flow in the exchange currents also increases. In Chapter 17 the future outflow at Gibraltar is estimated, and this complication is handled by a year-to-year step-wise numerical integration.

The hydrologic statistics of the Mediterranean are not known with precision but a fairly consistent model based on the assumption of a salt balance (not quite true) and information from several sources gives the following rates in cubic meters per second:

Evaporation losses..................35,000
Gibraltar Outflow................789,000
Gibraltar Inflow...................814,000
Modern river inflow................10,000

These inflows and outflows have been adjusted to balance, but the precision needed to balance this budget with measured values is lacking, and strictly speaking, the salinity is not now in a steady state.

What is the relation between the higher salinity of the Mediterranean and its outflow ? The answer to this question is of great interest to us because the outflow plays a very important role in conveyor-belt circulation and climate. The precise dependence of outflow on salinity is a complicated matter. However, by neglecting some smaller factors in the driving force in a simple approach, it appears that, to a reasonable approximation, the outflow is directly proportional to the density difference between the dense outgoing water and the less dense incoming Atlantic water at the point where the exchange currents cross the Gibraltar sill. Similarly, the outflow would be proportional to the salinity difference, after allowing for effects of temperature changes on the densities of the exchange currents.

We are intensely interested in the effect of the outflow on the amount of saline water reaching the Nordic and Labrador Seas, and any other high-latitude areas where deepwater might be formed. In that regard, one important thing now can be added to the discussion of the fate of the outflow in the preceding chapter. It is this: Although the volume of the outflow may be approximately proportional to the salinity (density) difference between inflow and outflow at the sill, the resulting effect on high-latitude sea-surface salinity is not at all proportional. That is, the effect must be highly nonlinear (Fig. 29). If the salinity difference at the sill is small enough, all of the low-density outflow would mix to buoyant equilibrium only in the upper few hundred meters of the Atlantic. It would then all go south and become diluted in the Gulf Stream, and none would reach the high latitudes by any direct path. If the difference is somewhat larger, some of the more-dense outflow would sink more deeply and move westward, forming Reid's wedge of slightly more saline middepth water that stretches across the mid-Atlantic. Some of it would diffuse up into the North Atlantic Drift and reach the high latitudes by way of the conventional conveyor-belt flow. If the difference increased to that of today, much of the deeper mixture might reach the Nordic Sea by the more direct and deeper path that Reid proposed for the outflow. Although some of the outflow is found to the south and west, the data reported by Schönfeld and Zahn suggest that a large part moves northwestward in the 1000-1500 m-depth zone. Eventually, by a poorly defined route, it makes up much of the more-saline mixture that is suggested by the eastern regime of higher flow velocity moving northward just west of Ireland in the analysis of Greatbatch and Xu. Beyond that point, it continues to rise and mix into surface water west of northern Scotland (Fig. 20).

With a more extreme salinity difference at Gibraltar, the outflow would be much more dense than today. The greater the density difference relative to the Atlantic water, the slower is the mixing. Therefore, a large part of a very dense outflow would be expected to sink too deeply into the Atlantic to be a part of the flow that reaches the surface west of Scotland. Consequently, the deeper part would not contribute salt to enhance the deepwater formation that is needed to drive the conveyor-belt circulation. With these considerations in mind, we can now turn to an explanation for the anomalous deglaciation that began at the maximum of the Saalian Glaciation in Europe (Illinoian in North America) and elevated world sealevel to +7.4 m during a Milankovitch insolation minimum.

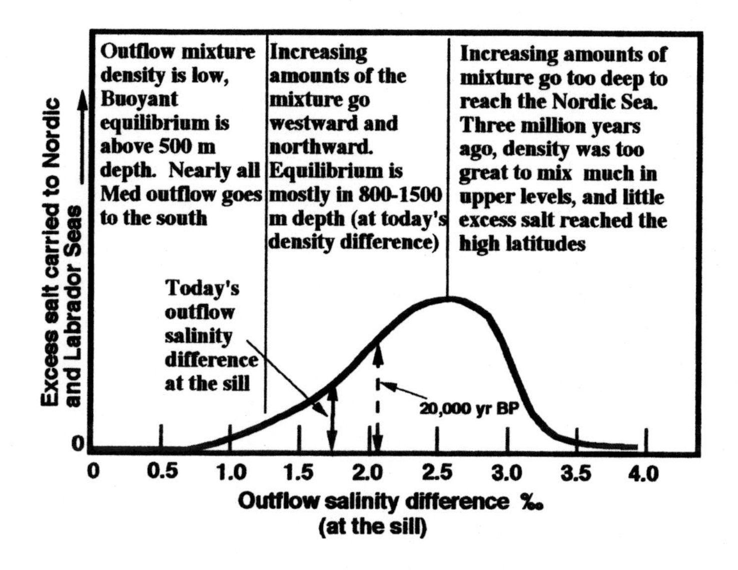

Figure 29: How Mediterranean salinity affects Nordic Sea salinity. Schematic nonlinear amount of excess salt reaching the higher latitudes in the North Atlantic plotted as a function of the salinity difference between the Mediterranean outflow and Atlantic inflow water at the Gibraltar sill. Although this curve is totally speculative, it is consistent with a progressive increase in the dimensions of the Gibraltar Strait and a resulting decrease in salinity of the Mediterranean outflow over the last four million years, and may explain the increase in the magnitude of the glacial cycles over the last 3 million years. The dashed-line arrow suggests the salinity at the ice age maximum when lower world-sealevel constricted the Gibraltar Strait. See Chapter 15.

11

How the Milankovitch hypothesis failed

In Europe and Asia, from 200,000 to 128,000 yr BP, the ice age that preceded our last ice age produced an ice sheet of enormous extent. This is depicted in Figure 30 from a 1995 paper by Arkhipov, Ehlers, Johnson, and Wright. Over an interval of 15,000 years after 160,000 yr BP, in the northern polar latitudes summer insolation was less than today while south of 65° N, in most of the ice sheet areas, the lower-latitude insolation was considerably greater. In accord with Milankovitch thinking, the energy balance of this insolation pattern should have caused deglaciation, as discussed in detail in Chapter 15. Instead, ice volume increased and most of the period after 160,000 yr BP was a time of unusual ice-sheet growth, notably in Eurasia where the record is preserved. In this interval the small ice sheets in Scandinavia and northern Siberia grew and merged to form a single massive ice sheet from Ireland eastward almost to the Lena River in eastern Siberia. All the rivers in Europe and western Siberia that drain into the Arctic Ocean became blocked by ice.

East of the Ural Mountains, most Siberian rivers were forced to discharge into large lakes that accumulated along the south edge of the great ice sheet. The summer-season meltwater from the southern edge of the ice sheet also accumulated in these lakes. The lake levels rose and they all merged into one vast freshwater inland sea bordering the ice sheet. Finally, the giant lake found an outlet to the Aral

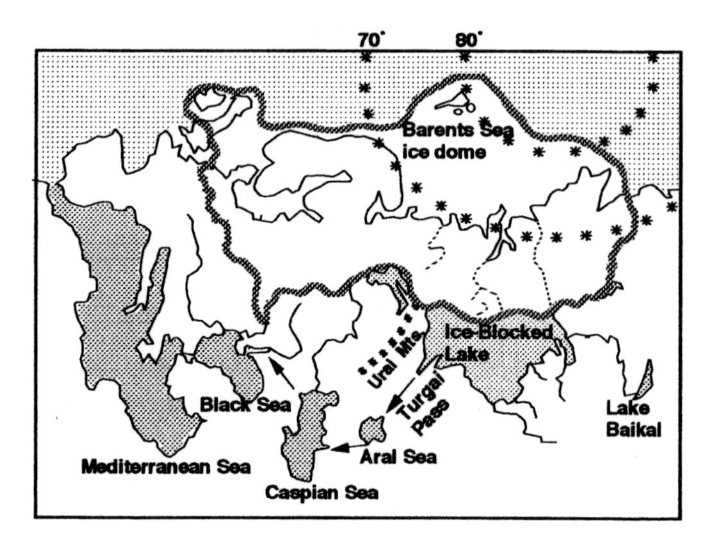

Figure 30: The Eurasian ice sheet about 144,000 yr BP. The giant ice-dammed lake was formed when most Siberian rivers to the Arctic were blocked by ice. The meltwater overflow into the Mediterranean eliminated the salty Mediterranean outflow and initiated the anomalous deglaciation by shutting down North Atlantic Deep Water formation. Map redrawn from Arkhipov et al. (Boreas, vol. 24, 1995). Used with permission of Taylor & Francis AS.

Sea over a low zone called the Turgai Pass. The overflow soon filled the Aral Sea, and, flowing over lowlands that had once been shallows of the ancient Tethys Sea, the overflow entered the Caspian Sea. The Caspian Sea had been receiving seasonal meltwater from the Volga River west of the Urals, but in the dry ice-age climate, evaporation losses had prevented it from becoming filled. However, with the addition of the Aral Sea overflow, the Caspian also quickly filled and poured over into the Black Sea. The Black Sea had been receiving seasonal meltwater from the Danube and Don rivers, but likewise had not been filled. But now the Black Sea was receiving meltwater from the entire 5000 km of the southern front of the great Eurasian ice sheet, and it became the last domino to fall. We do not have direct evidence for the meltwater outburst into the Mediterranean from the Black Sea, but what happened can be confidently inferred as a mirror image of a well documented similar event that occurred about 8000 yr BP. This is a fascinating story in itself.

The Black Sea is connected to the Mediterranean by the Bosporus Strait (Fig. 31), a narrow channel that is 26 km long and at some points only a kilometer wide. Its depth to bedrock is about 100 m today, although the water depth to the soft sediment above the bedrock at the south end is only about 35 m. The infilling sediment that nearly blocks the channel is a transient result of ice-age sealevel fluctuation. During interglacial times when the sealevel is high, and for a few thousand years after sealevel begins to fall, the channel at the south end becomes largely filled with sediment and gravel that is eroded by the Kagithane and Alibey rivers that drain through the Golden Horn embayment into the south end of the Bosporus channel. The sedimentary material deposited there is carried northward into the strait by the deeper and more saline exchange current, which flows from the Mediterranean toward the Black Sea. This process has filled the channel with sediment to a water depth of 35 m at the south end since about 8000 yr BP when an earlier sediment dam was washed out. Sealevel in the Black Sea 8000 years ago was 140 m below that of today, and the sediment dam of that time, with its top 19 m below present sealevel, was

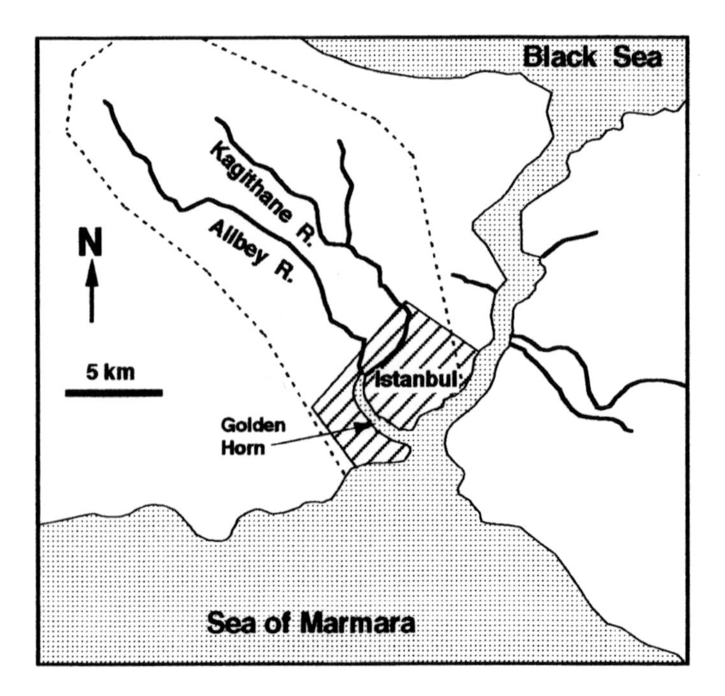

Figure 31: The Bosporus Strait between the Black Sea and the Sea of Marmara, which connects to the Mediterranean. The ~100 m-depth of the channel to bedrock is filled to a depth of about 35 m at the south end by sediment and gravel eroded by the Kagithane and Alibey Rivers and carried into the strait by the lower exchange current.

washed out to bedrock by the rising sea as the ice sheets of the last ice age melted and raised world sealevel. The story of the voyages to recover sediment cores from sites on the Black Sea continental shelf to define this event has been told in the book: *Noah's Flood* by Ryan and Pitman and in a journal paper by Ryan et al. in 1997.

One can imagine a critic asking why the soft sediment

plug that had earlier dammed the Bosporus was not eroded out by Black Sea outflow when the world sealevel began to fall below the top of the dam as the last ice age began. The short answer is that the Black Sea level also fell below the top of the dam not too long afterward. The following detailed explanation has interesting climatic implications.

During the wetter interglacial climate there is a generous rate of inflow of rivers into the Black Sea. This inflow produces a strong outflowing current through the Bosporus. During interglacial times there is also a much greater volume of deep saline Mediterranean water flowing northward into the south end of the Bosporus than flows into the Black Sea at the north end. This is because the stronger upper-level flow from the Black Sea mixes into and entrains much lower-level Mediterranean water and carries it back to the south. According to Gunnerson and Özturgut's discussion in 1974, the Mediterranean water that does finally reach the Black Sea today does so intermittently. As will be discussed in Chapter 12, the last ice age began in Canada, and the relatively moist interglacial climate in Eurasia continued for at least four thousand years. As sealevel fell when the last glaciation began, the constriction of the channel would have shut off any intermittent saline inflow to the Black Sea at the north end of the strait, although the saline inflow at the south end coninued. The river flow then slowly flushed out the salt.

A few thousand years after the last ice age began in Eurasia, when world sealevel was slightly below -19 m, the increasing aridity in the Black Sea region caused its sealevel to also fall below the top of the dam, thus preserving the dam and the freshwater content of the Black Sea until the last deglaciation. However, at the start of the previous ice-age cycle about 198,000 yr BP with a similar sediment dam in place, the world sealevel fall began with $\delta^{18}O$ values already more positive, suggesting that Eurasian glaciation was already in progress with an arid Black Sea regional climate. Consequently, on that earlier occasion the Black Sea level fell faster than world sealevel, leaving the Black Sea as salty as the Mediterranean.

The buildup and catastrophic washout of a sediment dam in the Bosporus has probably been a repetitive process with each ice-age cycle over the last 2-3 million years, and is the likely cause of the erosion of the bedrock in the Bosporus channel to its present depth. It is easy to infer that the washout of the Bosporus dam, which occurred from south to north about 8000 years ago, was duplicated in reverse from north to south about 144,000 yr BP when Siberian rivers were blocked and meltwater filled the Black Sea, topping the dam of that time and washing it out into the Sea of Marmara (Fig. 32). This happened at the maximum of that earlier ice age, known as the Saalian in Europe, with the Mediterranean sealevel about 120 m below present. This low sealevel (Fig. 8) is inferred from the marine $\delta^{18}O$ record of that time and the data of the analogous last glacial period obtained from cores in coral deposits off Barbados by Fairbanks (1990). The top of the Bosporus dam was probably about 65 m above the bedrock channel of the Dardanelles Strait at the outlet to the Sea of Marmara, which was about 35 m above the low full-glacial sealevel. Consequently, when the meltwater mixture washed out the Bosporus dam, the cataract through the Bosporus had an initial 65 m-fall, which was followed

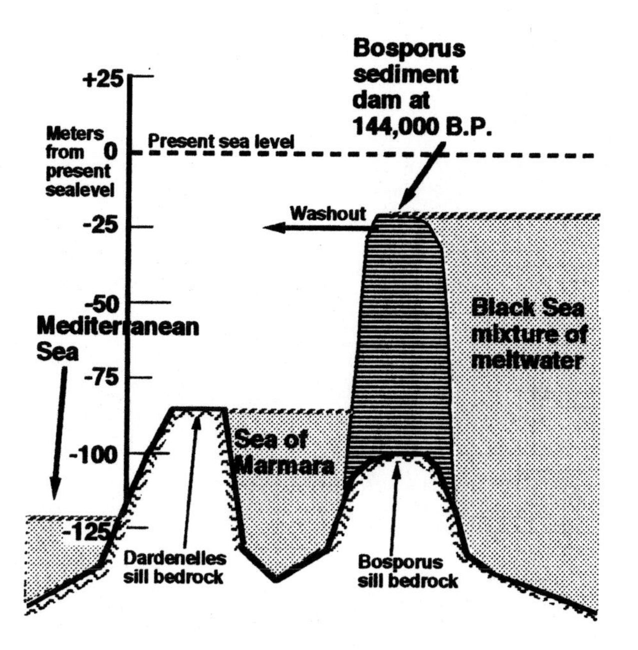

Figure 32: Washout of the Bosporus dam, 144,000 yr BP. Schematic diagram of the sediment dam in the Bosporus at the moment of the washout, when meltwater overflowed the Black Sea at the maximum of the Illinoian/Saalian glaciation.

by a 35 m fall through the Dardanelles. The Black Sea level probably fell about 60 m in two or three years.

This pulse of a mixture of saline meltwater from the Black Sea would have been equivalent to a layer 12 m-thick over the entire Mediterranean, and is about ten times the annual hydrologic deficit. The pulse would therefore have reversed the exchange currents at Gibraltar (Fig. 33), and no highly saline outflow would have entered the North Atlantic. The continuing drainage of meltwater into the Mediterranean from the entire southern front of the

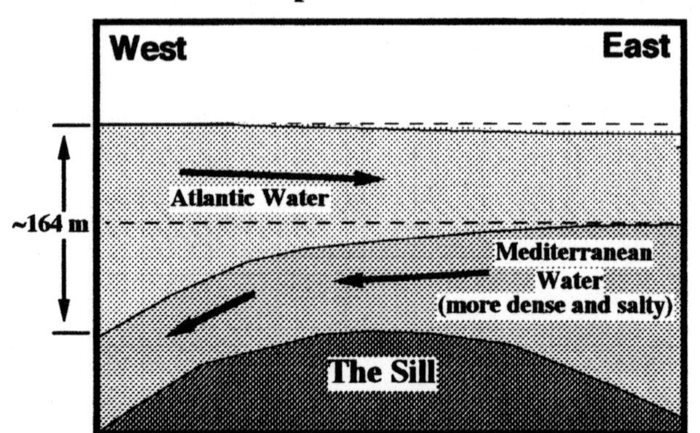

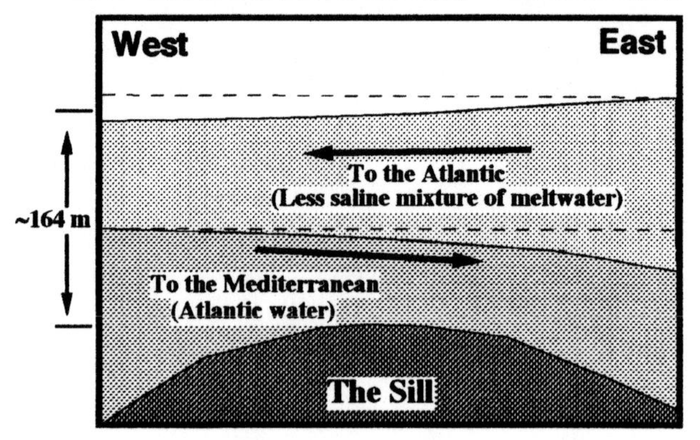

Figure 33: Exchange currents reversed by meltwater flooding. Currents over the sill in the Strait of Gibraltar before and after the washout of the sediment dam in the Bosporus 144,000 years ago.

Eurasian ice sheet is estimated to have equaled or exceeded the evaporation losses, which would have ensured the continued reversal of the exchange currents.

If this pulse of meltwater triggered Eurasian deglaciation in the manner argued below, the rate of meltwater discharge reaching the Mediterranean would have been almost three times larger than the Med evaporation losses, as estimated by analogy with the Mississippi River flow during the last deglaciation of the North American ice sheet. The exchange currents would have remained reversed and the salty outflow from Gibraltar would consequently have been totally eliminated for as long as the large meltwater supply lasted. During this time no significant deepwater formation could have occurred in the high-latitude North Atlantic.

We even have an inferred age for the onset of the meltwater influx into the Mediterranean. Thunell and Williams in 1982 published a detailed study of core TR172-22 from the eastern end of the Mediterranean (Fig. 27) in which they found an abrupt decrease in cold-water foraminifera and an increase in warm-water species at the beginning of the last interglacial (Fig. 34). Surface temperatures became suddenly warmer when the less dense meltwater mixture from the saline Black Sea of that time flowed into the eastern Mediterranean and stratified the upper layers, thus allowing summer warmth to accumulate in the surface. At the same time the diatom population greatly increased, implying a much larger increase in nutrients, as would be expected with an influx of meltwater. In 1982 the anomalous deglaciation was unknown, and Thunell and Williams correlated the thick layer of interglacial deposits with the world ocean $\delta^{18}O$ peak about 125,000 yr BP. To be consistent with the short duration of the world ocean $\delta^{18}O$ peaks, however, and to explain the large thickness of the layer of interglacial diatoms and foraminifera in the core sediment, they assumed a deposition rate for this layer that was three times the measured average rate from today to 125,000 yr BP. However, this assumption was unnecessary. In the Mediterranean there was no cold Antarctic Deep Water to deceitfully bias the $\delta^{18}O$ record in a positive direction, and the full effects of stratification and the negative $\delta^{18}O$ of the meltwater mixture from the Black Sea are revealed. The logic principle of Occam's Razor says that the simplest interpretation compatible with the evidence should be made. Therefore, we use the average deposition rate, correlate the warm species peak (Fig. 34) with the maximum Eemian warmth at 123,000 yr BP, and extrapolate to place the beginning of the anomalous deglaciation at 144,000 yr BP.

There is an important lesson to be learned from he sapropel of Figure 34. The early part of the sapropel, with a large deglaciation indicated in the Mediterranean sediments, occurred near an insolation minimum. Therefore, the practice of "tuning" that identifies sapropels with insolation peaks would cause age errors where other anomalous deglaciations have occurred, as is likely in the case of the cold-climate sapropel dated to 525,000 yr BP by Rossignol-Strick et al. in 1998.

Over the next few decades after the Bosporus dam broke, a most profound change would have occurred in the North Atlantic and shortly afterward in the Southern Ocean around Antarctica. In the Nordic Sea and other high-latitude areas the sea-surface salinity fell because, with no excess salt from the Mediterranean, the salinity became dominated by the low salinity of the surface water in the

Arctic Ocean. Consequently, deepwater formation no longer occurred anywhere in the North Atlantic. No replacement water from the Gulf Stream latitude flowed northward, the northern gyre became cold, and lower temperatures extended southward all across the north central North Atlantic.

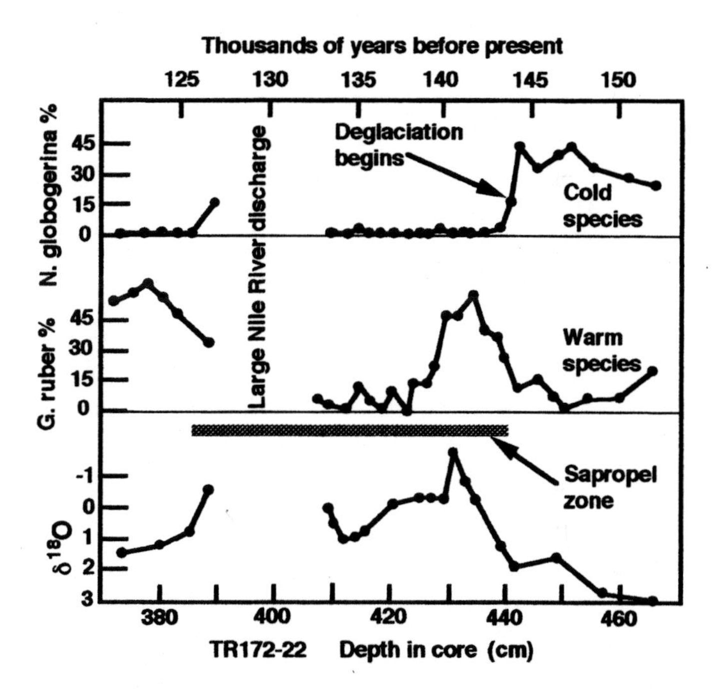

Figure 34: Meltwater effects in the eastern Mediterranean. Selected data from Thunell and Williams (1982) illustrating the change in foraminiferal species in the eastern Mediterranean when glacial meltwater from the Black Sea began to flood into the eastern Mediterranean about 144,000 years ago. Used with permission of Elsevier Life Sciences. The gap in the data centered about 130,000 years ago is the probable result of large Nile River discharges during an interval of strong monsoons in Africa. The salinity at that time was too low for the foraminiferal mollusks to survive in the eastern Mediterranean. The sapropel zone is characterized by abundant diatoms.

A paradox resulted because the oceanic cooling started the anomalous deglaciation by indirectly accelerating the rates of summer melting along the southern front of the great Eurasian Ice Sheet. Summer temperatures on land would have been dominated by the incoming solar energy, which was greatly increased at ground level by the clear skies associated with a smaller moisture supply to the ice-sheet regions. The climate of Eurasia and to a lesser extent the climate of northern North America was therefore switched abruptly from an ice age maximum to an active state of rapid deglaciation due to a reduced moisture supply. To see how this reduction occurred we need to look at the storm paths and the atmospheric flow over the North Atlantic.

The general atmospheric flow in the midlatitudes is approximated by the eastward flow of the prevailing westerlies. This river of air is whipped back and forth, north and south, by the jet streams that tend to follow along the borders between surfaces of land, water, or ice having contrasting temperatures, as discussed in Chapter 9. South of the jet stream, temperatures are generally warmer. Today, the low-pressure storm systems following the jet stream over the northeastern United States are steered northeastward by the contrast between the warm Gulf Stream south and east of Newfoundland and Greenland and the colder land masses to the west and north. During glacial times of ice-sheet growth and relatively less deepwater formation, this jet stream flow was not as strong as today because the northern gyre was not as warm. But the flow was still present to some extent, and the moisture that storms picked up when passing over the ocean all the way from the Gulf of Mexico to the sea west of Scandinavia was delivered as snow to the ice sheets in eastern Canada and Greenland or carried to Scandinavia and on into Siberia, feeding snow to those ice sheets along the way. This circulation pattern changed completely when winter-cooled water did not sink to form intermediate-level deepwater, and the gyre became cold. In the new pattern (Fig. 35), a

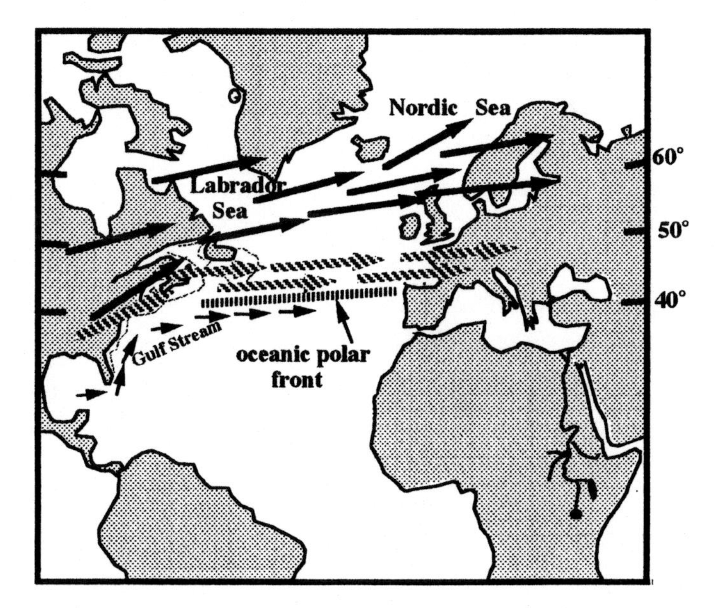

Figure 35: Winter storm tracks before and after Med outflow loss, when meltwater from the southern front of the Eurasian ice sheet flooded the Mediterranean. Before the flooding, storm tracks (black arrows) were broadly distributed over the northern North Atlantic. After the flooding the Atlantic north of latitude 40°N became very cold, and the winter storm track paths (hatched arrows) narrowed to a southerly zone along a strong zonal oceanic polar front near the Gulf Stream, thus depriving the ice sheets of snows. See text.

strong oceanic polar front with a large temperature contrast would have occurred from east to west across the Atlantic along the northern edge of the Gulf Stream at about the latitude of northern Spain. This probably became the main storm track over the North Atlantic. One can infer two contrasting seasonal atmospheric circulation patterns over western Eurasia that would have resulted from the cold northern North Atlantic. In winter, the temperature difference would have been small between cold European land surfaces and the ocean to the west, which was close to 0°C. The atmospheric flow would have remained quite zonal with flow from west to east, and very little moisture was carried northward to the Eurasian Ice Sheet. Winter temperatures in Europe would have been very cold.

In summer, incoming solar energy warms the land surface much more than the ocean, which has a large thermal inertia due to the 100-200 m-thick mixed layer of surface water. Consequently, the atmospheric jet stream flowing between the cold ocean and the much warmer European land surfaces would have steered storm paths northeastward over the land masses of western Europe from as far south as arid North Africa. Large amounts of relatively dry air would therefore have passed over the Eurasian Ice Sheet. The somewhat warmer and dryer air with minimal cloud cover would have greatly accelerated summer melting, and the anomalous deglaciation of the Eurasian Ice Sheet began. It was indeed a deglaciation under a dry climate.

The positive feedback of the accelerated melting along the southern Eurasian ice-sheet front kept the deglaciation going by maintaining the large meltwater input to the Mediterranean. The storm tracks associated with the zonal

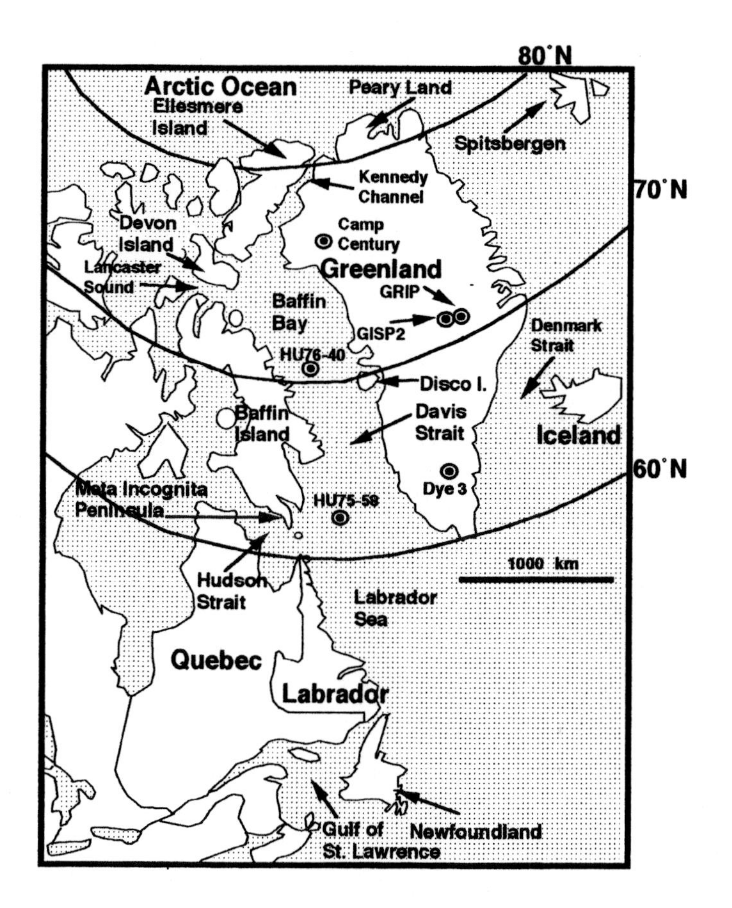

Figure 36: Baffin Bay and surrounding regions showing ice core and sediment core sites and locations mentioned in the text.

temperature contrast took moisture away from the Laurentide Ice Sheet in Canada, as storm paths moved out to sea farther south along the Gulf Stream. On Greenland, the ice-core record is clearly consistent with a large loss in the moisture supply to the ice cap. Koerner in 1989 reported that during the last interglacial, the glacial ice disappeared at the Dye 3 and Camp Century ice core sites (Fig. 36). Except for a broad band of ice remaining in central Greenland in the thickest part of the ice cap, Greenland to the north and south became ice-free. No similar loss of ice happened during the last deglaciation, which occurred largely with a conveyor-belt circulation like today that maintained Greenland's moisture supply.

The loss of deepwater formation in the North Atlantic would have been strongly felt in the Antarctic. With the loss of saline deepwater from the North, the Antarctic Convergence Zone (Fig. 21) surrounding the continent would have been pushed far to the north, just as reported in the CLIMAP studies of the glacial maximum near 18,000 BP, and a much larger area of perennial sea ice would probably have developed in only a century or two. The open-ocean source of moisture for precipitation on Antarctica therefore became much more distant. With the wider band of sea ice around the continent, the zone of ice-to-ocean temperature contrast moved northward, and the penetration of storm systems to the Antarctic ice sheet was much less frequent. Therefore, the continent became colder and precipitation on Antarctica would have decreased.

In cold Antarctica little surface melting occurs. Ice loss is controlled by the flow of ice off the continent into the sea, and this flow occurs all the way down to the base of the ice where basal sliding often occurs, and where, over much of the thick ice, the effect of colder surface temperatures would not be felt for a few thousand years. With less precipitation and with continued flow into the sea, the ice volume in the Antarctic would also have decreased. Deglaciation was probably also accelerated by the sealevel rise, which destabilized major ice shelves with inner edges standing in water on the fringes of the continent. Clark and colleagues argued convincingly in 1996 that an abrupt sealevel rise of 18 m about 14,000 yr BP was due to loss of Antarctic ice, which was probably destabilized in this way, with large tabular icebergs floating away and melting in the open ocean.

The unique world-ocean condition that was brought about by the total lack of deepwater formation in the North Atlantic ended about 6000 years later when the Eurasian Ice Sheet eventually retreated north of the divide in central Europe, and the Siberian rivers again flowed to the Arctic. Significant amounts of glacial meltwater no longer entered the eastern Mediterranean, and the saline outflow at Gibraltar resumed. But the most recent deglaciation required 11,000 years to completely remove northern ice, and in the 6000 years that ended about 137,000 yr BP, it is likely that only about half of North American and Eurasian ice was removed. The anomalous high sealevel found on Barbados probably lasted only 600-1000 years and the renewal of moderate deepwater formation merely restarted ice-sheet growth on the large masses of Northern Hemisphere residual ice. The restart began about 136,700 yr BP when sealevel began to fall below the oldest wavecut notch on Barbados made during the last interglacial. A relevant question here is: Why was there not a transition to strong conveyor-belt circulation and a warmer climate at some point during the anomalous deglaciation ? The answer will be easier to consider later in the context of the more conventional recent deglaciation, discussed in Chapter 13.

Note that the high sealevel at the end of the anomalous

deglaciation did not occur during a true "interglacial period." In Europe there was neither an interglacial climate nor a complete loss of ice like today. Ehlers, Meyer, and Stephan reported in 1984 that the Scandinavian Ice Sheet in northern Europe quickly re-advanced, apparently overriding buried ice that had not yet melted during the anomalous deglaciation. From a European point of view, one could argue that the anomalous high sealevel did not contradict the Milankovitch hypothesis with respect to climate temperatures and regional ice volume. On the other hand, I.J. Winograd's Devil's Hole data (Fig. 10) clearly indicate a warm interglacial climate in the southwest United States, as one might expect if the southern front of the North American glacial ice had undergone a major retreat.

As noted earlier, however, the subsequent sealevel fall to about 80 m below present from 136,700 to 130,000 yr BP during strong and rising insolation (Fig. 10) is a clear and inescapable Milankovitch contradiction. We know that on these two anomalous occasions after 144,000 yr BP, the orbital insolation factors were powerless to prevent major contrary changes in volumes of glacial ice. But we are making progress in understanding why the Milankovitch hypothesis failed at those times. The anomalous deglaciation demonstrates the strong effect on the North Atlantic circulation and climate caused by loss of all Mediterranean outflow. Conversely, an increase in Mediterranean outflow can explain the initiation of major new ice-sheet growth in Canada at the beginning of the last ice age, about 120,500 years ago.

Post press note: The Noah's Flood washout of the Bosporus dam has been challenged by Aksu et al. (GSA Today, May, 2002) on the basis of low salinity in the Marmara Sea due to presumed earlier Black Sea outflow. However, the 7600 BP ages of salt water mollusks that lived in a 1 cm-thick mud layer at the surface of the formerly dry Black Sea shelf (Ryan et al.) can only be explained by abrupt and violent Mediterranean inflow then.

12

Triggering the last ice age

At the end of the last warm interglacial period 120,000 years ago, a decline of about 0.3‰ began in the plotted $\delta^{18}O$ curve of benthic sediments. The decline implies increasing ice volumes or oceanic cooling or both at the start of the last ice age. However, the ice-volume record shown by sealevels on Barbados is quite clear. There, dated coral stratigraphy shows that world sealevel fell 8 m in the 3000-4000 years after 120,500 yr BP. The precisely calculated insolation curves for all northern latitudes fell below present at 120,000 yr BP and continued downward to a deep minimum at 115,000 yr BP. One cannot escape the implication that falling Northern Hemisphere insolation was responsible for initiating the last ice age, as discussed by R. Johnson in 1997. Our task now is to refine that approach and to construct a more detailed and precise model of the ice-age triggering process.

The onset of new ice-sheet growth occurred in northeastern Canada on Baffin Island and northern Quebec where today's summers are short and cold. On the higher elevations of Baffin Island large plateau areas are now covered by perennial snowbanks that seldom melt, as shown by photos taken by Earth Resources Technology Satellite 1 and discussed by Barry, Andrews, and Mahaffy in 1975. It seems that these areas are now at the threshold for ice-sheet growth. Similar conditions were probably present at 120,500 yr BP. If so and if the Milankovitch hypothesis were correct, the decrease of solar insolation to the 115,000 yr BP level in the Canadian area should have initiated large-scale glaciation, consistent with the known records of $\delta^{18}O$ and sealevel fall. With the advent of powerful computers, the virtual effect of minimum insolation at 115,000 yr BP was tested by Rind, Peteet, and Kukla using numerical models of the general atmospheric circulation. The results were reported in 1989. Even when significant increases in snowfall relative to the present day were allowed in the General Circulation Model, no ice sheet grew. They also tested an extreme initial condition: a 10 m layer of solid ice over the entire area. The model ice melted away and no lasting ice sheet was obtained. Either important factors were omitted in the model or the cooling due to reduced insolation was not the cause of the real-world renewal of glaciation.

When models fail, it is a good idea to look for answers in the published geological records where the data of interest may lie buried in the abundant earth-science literature, as in this case. Today on Devon Island at the north end of Baffin Bay (Fig. 36), there is a fairly thick glacial ice cap. It is a bit of the last ice age that did not melt away, although ice there had disappeared earlier during the anomalous deglaciation 140,000 years ago. An ice core was extracted from this glacier by Koerner, Bourgeois, and Fisher. In 1988 they reported that the layers of ice formed from snow deposited next to the bedrock at the very beginning of the last ice age had $\delta^{18}O$ values showing that the snowflakes of that time crystallized under relatively warmer and more humid conditions than was the case for layers higher in the core. In the basal layers they also found wind-blown pollen grains from willows. The firm conclusion is that, when the ice-sheet growth began 120,500 years ago, the first snows were accompanied by winds that blew from the south over warmer water in Baffin Bay and along a coast where willow trees grew. In this context in Baffin Bay, "warmer water" may simply mean that the bay was free of sea ice, even in winter. A heavy sea-ice cover insulates the air from the near-freezing water beneath the ice and allows air temperatures to fall far below freezing. With open water, air temperatures can be held close to 0°C. Today warmer water, that is, ice-free water, only occurs from June through October. Under conditions of ice-age initiation with a higher salinity, the bay may have been largely free of ice all of the year.

Warmer conditions are consistent with evidence from marine sediment core HU75-58 from 200 km off the southern coast of Baffin Island in the Labrador Sea, reported by Fillon in 1985 (Figs. 36 and 37). He measured the abundance of warm water foraminifera that are normally rare or absent off Baffin Island. There was a short interval when the shells of warm-water species were more than half the total population, just before the foram $\delta^{18}O$ dropped abruptly to cold ice-age values. The sedimentation rate was too low, however, to accurately measure the duration of this pulse of warm water.

The duration question has been answered by data from core MD95-2036 raised from the Bermuda Rise far to the south and reported by J. Adkins and colleagues in 1997. The sedimentation rate is much higher at this deep-sea location, which receives sediment carried by the deep-ocean currents from rivers emptying into the sea along the coast of northeastern Canada. In 1997 they reported relative amounts of particles of Canadian clay and iron oxide (hematite) as a function of depth in the core. At an estimated age close to 120,000 yr BP there was an abrupt six-fold increase in the abundance of clay and hematite in a narrow layer in the core. The entire jump occurred between adjacent data points about 100 years apart (Fig. 37). The high abundances lasted 400-500 years, and then settled back to a normal level. At this time the cold water at the site of HU76-40 farther north in Baffin Bay was slightly warmer, as reported in 1985 by A.E. Aksu.

What we see in this pulse of Canadian sediment is a 500 year interval of extremely high rainfall (and snowfall) with strong erosion in northeastern Canada. This implies an atmospheric circulation pattern quite different from that of today. The heavy rain and snow is precisely what is expected if the surface water in the northwestern Labrador Sea suddenly became warmer, as shown by Fillon's sudden high abundance of warm foraminifera. The entire 3300 km length of coasts (Fig. 36) from Newfoundland to Devon Island would have been bordered by water warmer and more ice-free than today. On the west side of Baffin Bay the lands would have been snow-covered most or all of the year. The large-scale temperature contrast would have attracted low-pressure storm systems to that area, and the cold high-pressure area over the high Greenland ice cap to the east would have tended to block eastward movement of such storms.

Consequently, an almost continuous pattern of heavy clouds and storms over northeastern Quebec and Baffin Island would have deluged those gloomy areas with snow throughout most of each year. With moisture from the

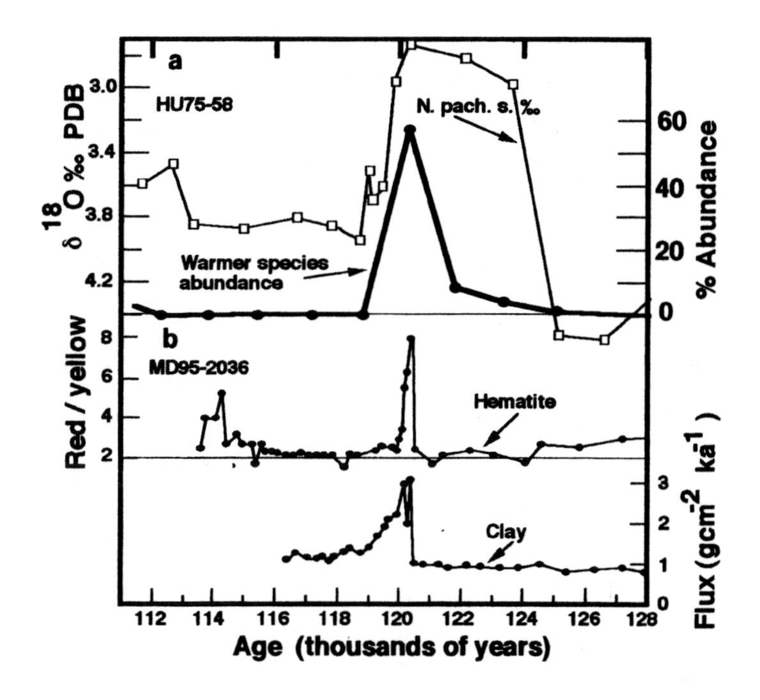

Figure 37: Climate-switch indicators as the last ice age began. Evidence for a much warmer Labrador Sea, with heavy rain and snowfall in northeastern Canada. The ages of the pulses are not measured precisely, and have been arbitrarily aligned at 120,500 yr BP, based on a consistent model of circulation changes at the start of new Canadian glaciation. Hematite and clay from the Bermuda Rise modified from Adkins et al., 1997). Species abundance and $\delta^{18}O$ modified from Fillon (1985) with permission.

Gulf Stream off Newfoundland entering this persistent storm system, torrential rains fell on the Maritime provinces and southern Labrador, and snowfall equivalent to a few meters of solid ice each year farther north would have been possible. Consequently, a rapidly growing ice sheet would have built up in the northerly areas over a few hundred years. As the ice thickened and the glacial surface rose, sea-surface temperatures fell in Baffin Bay due to regional cooling by the ice sheets and the icebergs that were discharged from the growing ice sheets through fjords around the bay. This cooling weakened the lock on the storm systems, and ended the period of extremely rapid ice-sheet growth on Baffin Island and Quebec almost as abruptly as it began. But the presence of the new massive ice sheets maintained strong temperature contrasts with water that was still warm in the Labrador Sea to the south, and ice-sheet growth in Canada and Baffin Island continued at a less spectacular rate for the next 3000 years or more.

In Europe the scenario was quite different. Detailed studies of quantities of tree pollen in ancient German lakes published by Field, Huntley, and Müller in 1994 show that there was a brief cooling in northern Germany about 120,000 yr BP that affected particularly the warmer-climate species (Fig. 38). The pollen of linden trees disappeared and did not return. Elm pollen disappeared and then returned briefly to a minor extent. Oak pollen nearly disappeared, but returned strongly for a time, finally diminishing and disappearing about 116,000 yr BP when pollen from all the warmth-loving plants disappeared. It seems clear that during the 500 year interval of catastrophic ice-sheet growth in Canada, northern Europe became cooler and dryer because the normal jet stream pattern across the northern North Atlantic was disrupted by the locked-in low-pressure systems over Labrador and Baffin Island. After about 500 cool years in Europe, a 3000-4000 year respite from the oncoming ice age began, although the

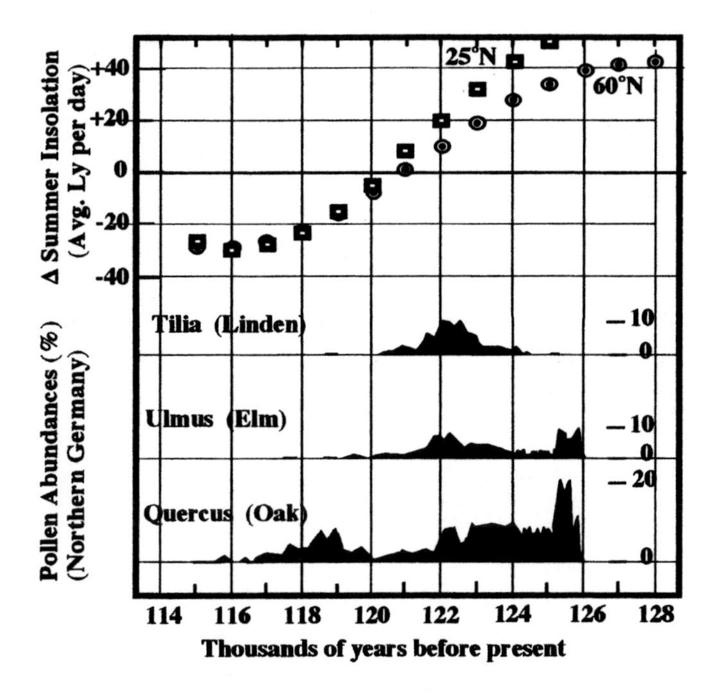

Figure 38: Tree pollen trends in Germany as insolation fell at 25°N and 60°N latitude. Pollen abundances of three warmth-loving tree species during the warm Eemian interval. The tree pollen nearly disappeared when the Canadian ice-sheet growth was initiated beginning about 120,500 yr BP. Only the oak species recovered. Insolation timescale is precise. Pollen timescale is less precise and minimal pollen probably occurs at the trigger point age for new Canadian glaciation. Pollen abundances modified from Field et al. (1994). Tabulated insolation from Berger (1978).

climate slowly deteriorated as the Canadian ice sheet expanded and cooled the high latitudes.

About 116,000 yr BP, however, a catastrophic event shut down the conveyor-belt circulation that had kept Europe warmer, and the ice age began there also. This event was a great "jokulhlaup," an Icelandic word for the release of water from an ice-dammed lake, in this case a dam formed by ice blockage of Hudson Strait. In 1976, Andrews and Mahaffy published results of a computer model of ice-sheet growth on Baffin Island, northern Quebec, and Labrador in which they chose 0.9 m of water as a maximum probable annual rate of precipitation. The model results suggest the possibility that after 8000 years of growth, the ice sheets on Quebec and Baffin Island could have expanded, come together, and blocked Hudson Strait, which after a few thousand years of ice-sheet buildup, was the only drainage channel to the sea for the entire central Canadian region surrounding Hudson Bay.

The lake that would have resulted was named Lake Zissaga (Fig. 39) by Adam in 1976, and the possibility of the collapse of such an ice dam about 116,000 yr BP was examined in detail by Johnson and Lauritzen in 1995. The York Canyons that cross Meta Incognita Peninsula bordering Hudson Strait were giant spillways for Lake Zissaga and its ancestral lakes over the last million years, whenever an ice dam formed in Hudson Strait. These spectacular canyons (Fig. 40) with spillway elevations about 300 m above sealevel are over 500 m-deep at their mouths on the north side of the peninsula, and were eroded into hard crystalline rock beneath the ice sheet by boulders scraped into the overflowing torrents in the canyons by the movements of the overlying glacier. It is likely that Lake Zissaga at 116,000 yr BP was dammed to the level of the spillways, and if so, the volume of water in the lake when the ice dam broke and the jokulhlaup occurred would have raised world sealevel by about 0.8 m. The presence of the

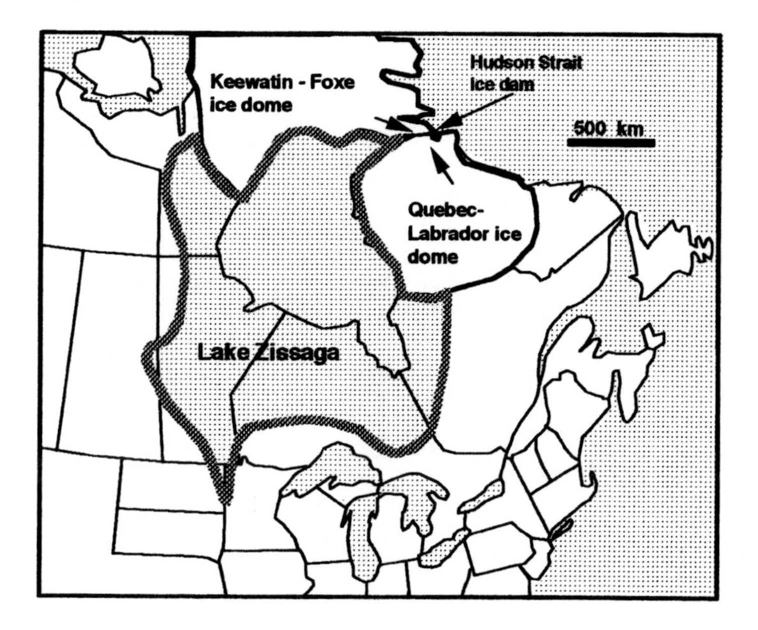

Figure 39: Lake Zissaga about 116,000 yr BP, about 4000 years after the Wisconsinan glaciation began. The full extent of the ice-dome boundaries is speculative. The boundary of the lake is consistent with the 300 m elevation of York Canyon spillways and the even higher strand lines on Baffin Island. Modified from Johnson and Lauritzen (1995) with permission from Elsevier Science.

dam at that time implies rain and snowfall rates averaging more than twice as high as the 0.9 m per year of the Andrews-Mahaffy computer model, because the ice dam was formed much sooner than predicted by the model.

The effect of the ice-dam collapse appears in the $\delta^{18}O$ curves from planktonic foraminifera dwelling in surface waters of the North Atlantic. The curves have a notch

Figure 40: York Canyon spillway for the Hudson Strait ice dam. The canyon was eroded into the hard crystalline rock across Meta Incognita Peninsula on Baffin Island (Blake, 1966). This canyon and one other were giant spillways for the overflow of the paleolakes that were impounded by ancient ice dams in Hudson Strait. At the distant end looking to the north, the canyon is over 500 m-deep. Air photo courtesy Natural Resources Canada, and reproduced with permission of Minister of Public Works and Government Services, 2001, and Courtesy of the Geological Survey of Canada.

implying colder surface water about 116,000 yr BP in cores V30-97, V29-179, and M12392-1, as reported by Ruddiman and McIntyre in 1981. A similar cold pulse was reported in the $\delta^{18}O$ of Norwegian speleothems by S.-E. Lauritzen in 1995. In the ice cores at the Camp Century and GRIP sites on Greenland, a brief cold spike

Figure 41: A debris-filled notch formed 116,000 yr BP. These small fossil coral cobbles sit on a shelf in solid limestone at the University of West Indies hillside site, and fill a wavecut notch in the last interglacial forereef slope. The notch was eroded when Lake Zissaga poured into the northern Atlantic as the Hudson Strait ice-dam collapsed. This briefly shut down deepwater formation and caused a small deglaciation and sealevel rise. A cobble of *Acropora palmata* coral from the hole at the top of the survey rod was dated to 117,000 ±1000 yr BP by C.D. Gallup (1994). Photo by R.G. Johnson.

occurs at this time in the oxygen-isotope ratios discussed by S.J. Johnsen and colleagues in 1995 and 1997. This is consistent with complete shutdown of the conveyor-belt.

On Barbados a debris-filled notch (Fig. 41) was formed as sealevel stopped falling and rose when the ice dam broke. This notch was cut into the forereef hill-slope of the last interglacial terrace at a world sealevel 8 m below the sealevel when the last ice age began. A fossil coral cobble in the notch was dated by C. Gallup et al. in 1994 to 117,000 ±1000 yr BP, which, within the statistical uncertainty of measurement, is the same as the age suggested by other lines of evidence. The freshwater flood into the Atlantic would have caused a more-zonal atmospheric circulation, and the deglacial effect on the ice sheets of that time lasted long enough to raise sealevel by 3.8 m (Fig. 16) as measured by R. Johnson at the Gibbons and Maxwell Hill sites on Barbados in 1998. The 3.8 m is the surveyed elevation difference between the floor of the debris-filled notch and a sharp secondary notch formed above at the subsequent brief sealevel maximum. Some of the rise may have occurred due to thick Laurentide ice that had flowed into Lake Zissaga, equilibrated at or above the 300 m lake level, and then collapsed when the lake level fell to sealevel. Destabilization of large accumulations of marine ice sheets grounded below sealevel is also a likely possibility. The duration of the jokulhlaup and its aftereffects, including the deglaciation causing the 3.8 m sealevel rise, could have been several hundred years.

The loss of conveyor-belt circulation is consistent with the total and lasting disappearance of pollen from temperate climate species of plants in northern Europe (Fig. 38). When the Hudson Strait ice dam broke, giant Lake Zissaga, the largest known lake in the world, flooded the North Atlantic with freshwater, stratified it, and turned off the conveyor-belt oceanic circulation, plunging Europe into the last ice age. On a smaller scale, this event was an analog of the anomalous flooding of the North Atlantic with lower salinity water from the Mediterranean that caused the conveyor-belt shutdown that began 144,000 yr BP (Chapter 11). Beginning about 116,000 yr BP, there was a similar partial deglaciation, a similar incomplete recovery of conveyor-belt circulation, followed by a similar rapid reglaciation. The conveyor-belt did not fully recover until 100,000 years later during the most recent deglaciation.

These events that occurred near the start of the last ice age are clearly visible in the records of pollen, ice cores, and marine sediments. The changes in atmospheric and oceanic circulation at that time are inferred quite logically from the evidence. The key factor underlying the sequence of events triggering major ice-sheet growth is the warming of the Labrador Sea. The cause of this warming is speculative, but there is strong evidence upon which to base a good working hypothesis.

In science, a working hypothesis is an idea that often explains data in a plausible or elegant way, but may not be convincing to all skeptics. A good working hypothesis will also lead to predictions that can be tested by new data or by calculations. In this case the hypothesis elaborated in the remainder of this chapter involves a lower sealevel in the Arctic Ocean, and is relevant for the initiation of the next ice age as well as the last. The underlying indirect cause for a lower Arctic sealevel was an increasing amount of excess salt in the high-latitude North Atlantic. The source of the rising salinity was the increasing outflow from the Mediterranean Sea, and there is an orbital cause for that increase. Increased outflow and decreased outflow are associated with opposite phases of the precession effect, when summers occur either far from or close to the sun.

Let's consider the cause of a decreasing outflow. The cause is the extension of African monsoons far to the north, resulting in a lower Mediterranean salinity. In 1983 and in

1985 Rossignol-Strick published results of a study that correlated northern precessional insolation maxima at low latitudes with sapropels formed in the Eastern Mediterranean Sea. Sapropels are dark-colored deep-sea sediment layers that have more than 2% organic carbon material. The Rossignol-Strick report identified strong Mediterranean sapropels with most of the insolation maxima over the last 300,000 years, and more recently Kallel et al. have found weak sapropels at times of most of the remaining maxima. In low latitudes, variations of summer insolation are due mainly to the precession effect, with its maximum every 23,000 years, more or less. These correlations are shown in Figure 42 together with the $\delta^{18}O$ from the SPECMAP marine timescale of Martinson et al.

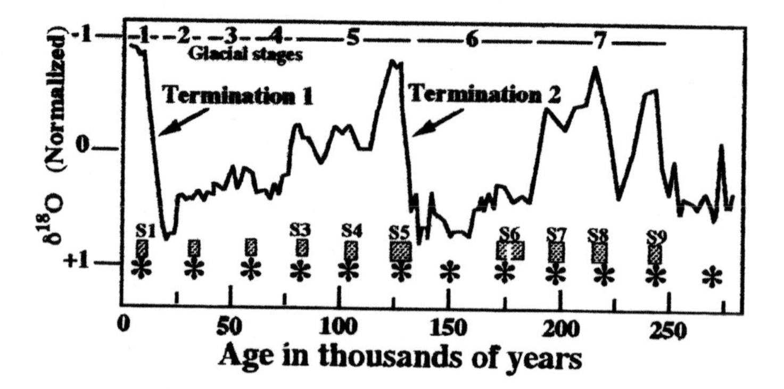

Figure 42: Sapropels, insolation maxima, and ice volumes. The Martinson et al. SPECMAP marine oxygen-isotope record (solid line) compared with Mediterranean sapropel ages (Rossignol-Strick, 1985, and Kallel et al., 2000) that correlate with tabulated 25°N insolation maxima shown by asterisks (Berger, 1978). Note that S5 actually extends from 125,000 to about 144,000 yr BP, see Figure 34.

Mediterranean sapropel layers form in the abyssal depths when the upper levels of the sea are more stratified and a good supply of nutrients is present in surface water to promote the growth of microfauna in the food chain. Sea life is abundant then and organic remains of sea creatures make up a larger fraction of the sediment that rains down to the bottom from the productive surface waters. When the surface salinity and density become slightly less, stratification occurs. This reduces the amount of winter convection between the surface and the deepest water, and diminishes the oxygen supply to the deeper benthic levels. The rate of oxidation of the carbon in the organic sediments is reduced and more organic debris is preserved, thus forming the black sapropel layers.

Figure 42 is one of the most significant illustrations in the book because the correlation of sapropel formation with insolation maxima implies the possibility that the Mediterranean is the critical link between the orbital precession effect and atmospheric-oceanic circulation changes in the North Atlantic that govern volume variations of northern glacial ice. Sapropel deposition is favored when African monsoons are quite strong, which is expected when strong summer insolation warms the northern subtropics. Factors affecting monsoons are discussed further in Chapter 15. When monsoons are strong, heavy rains extend far northward in Africa. The Nile River flow increases by a large amount. A flood of eroded mineral nutrients is carried into the Mediterranean by the Nile, and the Med salinity becomes less. In addition to the effects of the Nile discharge, atmospheric moisture mixes northward to the Mediterranean, making the atmosphere more humid and reducing evaporation losses, thus helping to make the surface salinity lower. Evidence for this description of conditions is found in the African environmental records reported by Yan and Petit-Maire in 1996. The upper Nile region from Aswan to the south was quite well watered at the beginning of the Holocene period about 10,000 years ago near an insolation maximum, and Nile flow was large. The lower-salinity conditions extended into the western Mediterranean as noted by Muerdter in 1984, and sapropel formation continued until about 7000 yr BP.

Consider first the effects of strong monsoons. During the insolation maximum about 10,000 yr BP, summer heating of the North African and Middle East desert areas was also at a maximum, and the strong atmospheric convection that resulted would have been a favorable factor driving the monsoons. Somewhat similar conditions occurred almost periodically about every 23,000 years and resulted in the sapropels of the Eastern Mediterranean. On such occasions, the salinity would have been lower in the entire Mediterranean. Consequently, the density difference between Atlantic water and Med water was less, and the outflow at Gibraltar was less. Pursuing this logic further, the high-latitude North Atlantic salinity would have been less and it was likely that little or no deepwater formed in the Labrador Sea. Without replacement water flowing to the Labrador Sea, the saline West Greenland Current would have been weaker, the Labrador Sea would have contained much sea ice, and there would have been a smaller land-sea temperature contrast. Consequently, traveling storm systems would have taken more southerly paths and tended to avoid the Laurentide Ice Sheet. Thus, cloud-free and sun-drenched summer conditions over eastern Canada can be inferred, favoring the deglaciation of Laurentide ice on these occasions. The general rule is that a small Mediterranean outflow favors a cold Labrador Sea, which promotes deglaciation by moving storm paths to the southeast and away from northeastern Canada, and by reducing water vapor transported northward from low latitudes into the atmospheric stream passing over Eurasia.

Next consider the effect of weak monsoons. The low summer insolation in the northern subtropics and the weak monsoons are the other side of the coin. There is then only a small Nile river flow, an arid climate over the Med, a more saline Med surface, and a stronger and more saline Mediterranean outflow. Voila ! That could be the source of the high salinity needed to increase deepwater formation in the Labrador Sea area, thus bringing in warmer and more-saline replacement water and initiating the last ice age. But this hypothesis has a problem ! How could the Labrador Sea have been warmed enough to bring in the abundant warmer-water foraminifera while the frigid low-salinity Canadian Current was flowing from the Arctic Ocean into Baffin Bay and southward down the Labrador coast ? And where in that region was the deepwater forming ? If the sea at the HU75-58 site off the south end of Baffin Island was warm enough for subpolar foraminifera to thrive, it was certainly too warm for deepwater formation there. Clearly this working hypothesis stands in need of repair.

In a journal paper, peer review would compel the author to stop and say that the answers to these questions are unknown. But this is not a journal paper, and disciplined analysis can carry us farther. Therefore, we will modify the working hypothesis, and move ahead to look for answers that are consistent with Fillon's record of warmer subpolar foraminifera off southern Baffin Island and Aksu's record of less extreme warmth at the HU76-40 site in Baffin Bay, 1000 km to the north.

Consider the critical question of deepwater formation. It could not have been forming off southern Baffin Island because the sea surface was too warm. The alternative is deepwater formation much farther to the north in Baffin Bay. Note that the distance in Baffin Bay from the south end of Baffin Island to the east of Devon Island is 1700 km, the same distance as from the south coast of Iceland to the south coast of Spitsbergen. To the west of Greenland there is the 2300 m-deep Baffin basin separated from waters to the south by the shallow sill (Fig. 36) at Davis Strait. To the east of Greenland there is the deep Nordic Sea basin, isolated from waters to the south by the shallower Denmark strait west of Iceland and Iceland-Scotland sill. One might expect winter-cooled water to sink and form deepwater in both these ocean basins at similar latitudes, particularly with favorable colder temperatures west of Greenland. However, today it only forms to the east in the Nordic Sea. The difference, of course, is the stronger flow of saline water over the sill into the Nordic Sea. The West Greenland Current with its weak salinity can maintain a strip of ice-free water hardly farther north in the winter than Disco Island. Its higher salinity is neutralized by the strong low-salinity Canadian Current coming into Baffin Bay from the Arctic Ocean.

The Canadian Current is driven by a slightly higher sealevel in the western Arctic Ocean that forces low-salinity water through the Queen Elizabeth Islands into the north end of Baffin Bay. There was dramatic evidence of this flow as the Arctic cooled in the fall of 2001, as shown by the GOES satellite images of sea-ice coverage provided by the NASA polar orbiter program. Between October 22 and November 1 the heavy sea-ice cover rapidly expanded from the broad mouth of Lancaster Sound between Devon Island and Baffin Island (Fig. 36). The open water east of the sound soon disappeared, and by November 1 the sea-ice cover extended most of the distance across the bay to Greenland. The ice subsequently expanded southward with the flow along the Baffin coast and eventually covered most of Baffin Bay, leaving open water only next to the Greenland coast south of Davis Strait, where the warmer and more saline West Greenland Current dominates the sea-surface conditions. Deepwater cannot form in Baffin Bay with this low-salinity water entering from the Arctic.

If deepwater formation could be started in the northern part of Baffin Bay, the dense deepwater formed would flow back toward the south over the Davis Strait sill, and the more saline West Greenland Current on the surface would increase to supply the replacement water, analogous to the Norwegian Current today. The warm sea surface southeast of Meta Incognita Peninsula and south of Davis Strait would be maintained, and Baffin Bay to the north would remain ice-free and stay cold enough for deepwater to form. Conditions for the ice-age initiation in Canada would then be met. Our disciplined search for the ice-age trigger therefore requires that: (1) The cold Canadian Current must be reduced or eliminated, or (2) its salinity must be increased enough to allow deepwater formation to begin in northern Baffin Bay.

It appears that these two requirements could both be met by eliminating most of the Arctic Ocean packice cover. Today, the existence of the packice depends on a layer of low-salinity water less than 200 m-thick that is maintained (despite the salt carried in by the Norwegian Current) by freshwater delivered by the rivers that discharge into the Arctic Ocean. This stratification prevents mixing of the surface water with the slightly warmer water below, and makes it possible for the surface to freeze during the winter. The low temperature of the packice surface during much of the year makes the polar area a region of high atmospheric pressure. However, the stratification could be destroyed by an increase in salinity of the Norwegian Current. This would increase convective mixing with the somewhat warmer water from deeper levels, and packice would be unable to form. The loss of packice would keep the winter atmosphere warmer and make the polar region an area of relatively low pressure, with more-frequent storms carrying water vapor out of the Arctic basin.

The effect of the loss of stratification and the resulting evaporation losses would be two-fold: (1) The Arctic Ocean sealevel would fall somewhat, with the water lost by evaporation being carried to surrounding continental areas and deposited as snow on growing ice sheets. (2) The sea-surface salinity would rise and reinforce an initial salinity increase that could have started the packice loss. Consequently, a weaker and more-saline Canadian Current could have combined with the stronger West Greenland Current to enable deepwater to form in Baffin Bay.

However, there is no evidence for a recent open sea in the sediment cores from the Arctic Ocean. But the sediment deposition rates are extremely low, about one millimeter in 500 years as reported for core T3.67-12 by Y. Herman in 1974. The probable interval of ice-free sea is only 400-500 years, and an altered sediment characteristic would be difficult or impossible to detect in such a thin layer.

There is a secondary effect that might also contribute to a lower sealevel in the Arctic Ocean. The Arctic sealevel must be influenced by the flow of the Norwegian Current into the Arctic and by the flow of the East Greenland Current out of the Arctic along the Greenland coast. Today, the wind-driven East Greenland Current outflow carries a 300 km-wide band of heavy sea ice during much of the year. If this band of sea ice were no longer present due to the higher Arctic salinity, the temperature contrast would be sharper between the cold Greenland ice cap and the open sea to the east. The traveling low pressure storms along the East Greenland coast would intensify, and the stronger north winds could strengthen the current along the coast and might draw water out of the Arctic at a more rapid rate, thus lowering the sealevel there.

It therefore is likely that the loss of much of the Arctic Ocean packice would have caused the onset of half a millennium of Labrador Sea warmth, deepwater formation in Baffin Bay, and the catastrophic growth of the early Laurentide Ice Sheet. These events would have been started by the increased outflow of the Mediterranean, which added salt to the Norwegian Current and tipped the balance in the Arctic from a stable sea-ice cover to an Arctic Ocean with much more open water. The sequence of events is summarized in block form in Figure 43.

In the Caribbean, the falling Milankovitch summer insolation would have kept sea-surface temperatures lower and evaporation losses to a minimum. Therefore, the salinity of the Gulf Stream would not have been rising, and there was no source to provide additional salt to the Arctic Ocean other than the increasing Mediterranean outflow.

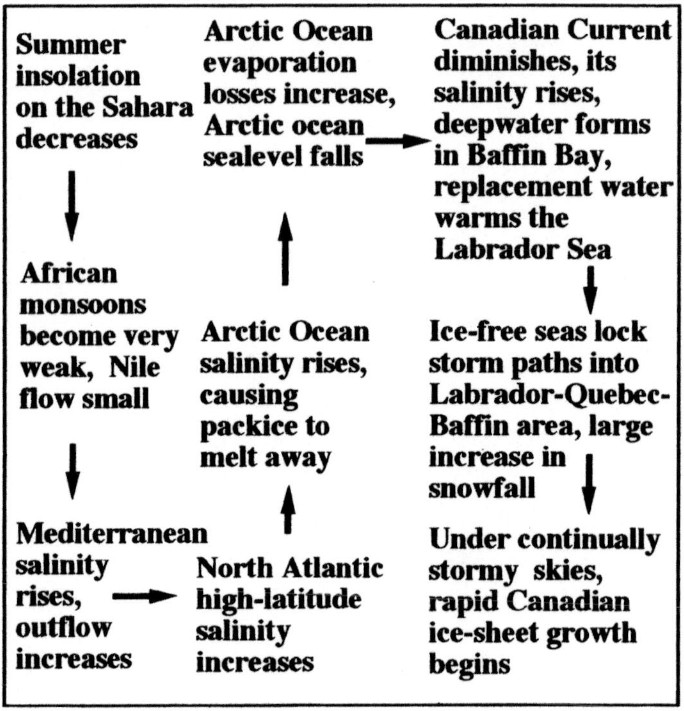

Summer insolation on the Sahara decreases

↓

African monsoons become very weak, Nile flow small

↓

Mediterranean salinity rises, outflow increases →

North Atlantic high-latitude salinity increases

↑

Arctic Ocean salinity rises, causing packice to melt away

↑

Arctic Ocean evaporation losses increase, Arctic ocean sealevel falls →

Canadian Current diminishes, its salinity rises, deepwater forms in Baffin Bay, replacement water warms the Labrador Sea

↓

Ice-free seas lock storm paths into Labrador-Quebec-Baffin area, large increase in snowfall

↓

Under continually stormy skies, rapid Canadian ice-sheet growth begins

Figure 43: The probable sequence of events that initiated new ice-sheet growth in Canada at the beginning of the Wisconsinan Glaciation 120,500 years ago. The initial interval of rapid ice-sheet growth lasted about 500 years.

13

Terminating the last ice age

Our quest for ice-age answers would not be complete without examining events of the last deglaciation and constructing a scenario to explain how the present warm interglacial climate was brought about after more than a hundred thousand years of ice-age climate. To get started, a brief review is in order.

As discussed in earlier chapters, we know that the present warm climate around the North Atlantic depends on a strong northward conveyor-belt flow of warm saline water to the vicinity of Iceland and the Nordic Sea (Fig. 20). This large flow occurs because in the winter the saltier water in the Nordic and Labrador Seas cools and sinks to deeper levels, forming deepwater that eventually flows to the Southern Hemisphere. The replacement surface water from low latitudes brings with it a supply of excess salt that allows the density to remain high to continue the sinking process. During the ice age the conveyor-belt was weak because a smaller amount of deepwater formed.

The warm conveyor-belt may have been the strongest deglacial factor in the last deglaciation, but the anomalous deglaciation under a dry climate gives us a greater insight into the details. The last deglaciation was caused by two critical effects that did not occur at the same time: (1) The atmospheric flow over the North Atlantic became more zonal and carried less moisture to the great ice sheets, thus allowing summer melting to dominate over annual snow accumulation. (2) The ocean and atmospheric circulation switched to a mode of strong conveyor-belt flow, thus warming the higher latitudes and accelerating the rate of summer melting. Without the switch to the conveyor-belt mode, no true warm interglacial climate could occur, as noted earlier in the case of the anomalous deglaciation.

The question is: How did the northern gyre become salty enough to get the strong conveyor-belt going ? This is not a simple question to answer because the tropical air flowing north to feed moisture to the ice sheets and make them grow likewise brought much rainfall and snow to the sea surface around Iceland. This kept the northern sea-surface salinity lower and apparently prevented the startup of the strong conveyor-belt flow during most of the ice age. Therefore, what is needed is a way to change the atmospheric circulation and temporarily minimize precipitation on the northern gyre. Once the strong conveyor-belt flow begins, it maintains lower-level deepwater formation, despite heavier precipitation, by bringing much excess salt northward from the Gulf Stream. Only a very large addition of freshwater to the northern North Atlantic can switch off the conveyor-belt, as in the examples of the Lake Zissaga jokulhlaup, when enough fresh water flooded into the Atlantic to raise sealevel 0.8 m.

This unpublished working hypothesis of proposed circulation changes that accomplished the last deglaciation is based on temperatures and salinities in the northern North Atlantic estimated by Duplessy and his colleagues, and published in 1992. The hypothesis is under current development by Johnson, Duplessy, Teller, and Wright.

Deglaciation was occurring rapidly 10,000 years ago, with the conveyor-belt warm-climate oceanic circulation locked securely in place by northward transport of excess salt. Northern Hemisphere insolation attained a maximum at that time. From a Milankovitch point of view, this is as it should be. But on many other occasions with equally high insolation, at 84,000 yr BP for example (Fig. 8), no strong and lasting conveyor-belt circulation occurred. Sealevel rose only to about 14 m below present shortly after 84,000 yr BP, and glaciation soon resumed. The Milankovitch hypothesis fails to explain these occasions. In scientific detective work, the first question one asks when confronted with such a problem is: What was different on these occasions, and how could the differences have brought about a warm interglacial 10,000 years ago but not 84,000 years ago ? There are two aspects of the Laurentide Ice Sheet that were quite different at the beginning of the last deglaciation: (1) The ice sheet had grown to enormous size and extended far to the south of the Great Lakes. (2) A partial collapse of the Laurentide Ice Sheet resulted in a large release of icebergs into the North Atlantic. These two factors were the keys to the transition to a true interglacial climate, because they made possible a series of events that led to a high salinity in the northern gyre and the Nordic Sea.

Relative to today, the world ice-volume increase before the insolation maximum about 84,000 yr BP was probably only about 40 m sealevel equivalent, and much of that was probably Eurasian ice. Therefore, the Laurentide ice volume was small and its southward extension then was also quite small. How different it was 18,000 years ago ! Then the Laurentide Ice Sheet reached almost to the Ohio River in the eastern United States (Fig. 1). It covered the entire Great Lakes area with ice of considerable thickness and extended across the Canadian prairies to merge with the ice sheet covering the Canadian Rocky Mountains. The Laurentide ice-volume, if melted, was estimated by Denton and Hughes in 1981 to cause a sealevel rise of 75 m.

The deglaciation began with a large surge of icebergs through Hudson Strait and out across the North Atlantic about 18,000 yr BP. This event, known as Heinrich event H1 (Fig. 44), consisted of two parts and lasted almost 3000 years. It cooled the northern seas and diverted storm paths away from the ice sheets. The first samples of glacial debris, melted from icebergs and deposited on the seabed all across the North Atlantic, were analysed by W.F. Ruddiman in 1977. Subsequently in 1988, H. Heinrich reported results from longer sediment cores showing that there were many intervals of rapid deposition of ice-rafted debris, from which it was inferred that event H1 was only the latest interval when great armadas of icebergs had floated across the Atlantic from Canada. No iceberg event of this magnitude occurred 84,000 years ago. The basic argument is that, as Heinrich event H1 ended, the large southerly extension of the ice sheet continued to supply the necessary midlatitude meltwater to maintain the zonal circulation that reduced precipitation on the Nordic Sea and enabled the switch to strong conveyor-belt flow.

A critical question relevant to this argument is: Why didn't the Laurentide ice extend southward into warmer latitudes until the equilibrium melting rate became high enough to equal the accumulation rate, resulting in an indefinitely-long ice-age maximum ? Such an equilibrium may actually have occurred briefly, because the southern front of the ice south of the Great Lakes was relatively

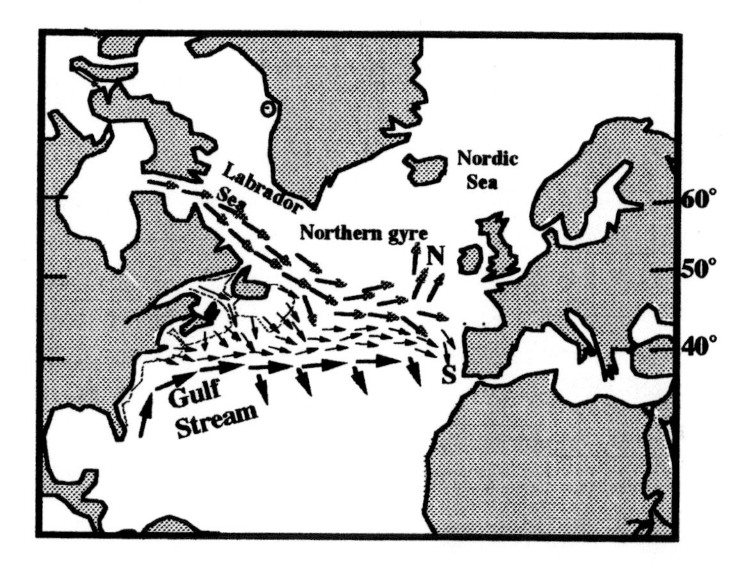

Figure 44: Iceberg paths during Heinrich event H1. During the first part of the event (patterned arrowheads), nearly all icebergs came from the central Laurentide Ice Sheet via Hudson Strait. During the second part of the event (small solid arrows), the main source was calving from ice sheets with their bases below sealevel on continental shelves in the Maritime Province area, New England, and Newfoundland. N - core NA87-22, S - core SU81-18.

stable before deglaciation began about 18,000 yr BP. But Heinrich event H1 is the answer to the question. It interrupted the last ice age by cooling the northern North Atlantic (see also Chapter 15 for amplification of cooling) and thus steering major storms eastward instead of northeastward over the Atlantic and eastern parts of North America. Therefore, event H1 accomplished much more than contributing its own icebergs to the deglacial process. It initiated the 11,000 year interval of deglaciation in North America and Eurasia by reducing the moisture supply to both American and Eurasian ice sheets. In addition, there was an important complementary deglacial process, caused by the deglacial rise in sealevel that can destroy ice sheets based on bedrock below sealevel.

During the interval of the glacial maximum before 18,000 yr BP, large ice sheets in the Northern Hemisphere and the Antarctic accumulated and flowed into shallow water on the continental shelves, with the base of the ice below sealevel. Their advance would have ceased when the rate of loss of ice by icebergs calving off into the sea equaled the rate of gain due to ice flowing out to the grounded edge. If the edge of the ice sheet is not stabilized by attached floating ice shelves that can occur when climate is very cold, the depth of water in which the ice can stand on the bottom is quite limited by the buoyant forces that float the ice and cause its more rapid breakup if the water is too deep. A small sealevel rise can disturb this equilibrium by increasing the buoyant forces at the outer edge and thus increasing the rate of calving breakup. A faster ice flow into the sea then results all the way from the toe of the ice sheet back into its higher elevations as the grounded toe of the ice moves back toward a new equilibration point.

In this process, the large flux of icebergs entering the ocean adds to the rise in sealevel, thus amplifying the effect of the original sealevel rise. Thus, a single small pulse of sealevel rise can result in a continuing deglaciation and a rising sealevel. For that reason the large marine-based ice sheets tend to shrink when sealevel rises, and conversely, tend to advance and grow when sealevel falls. An example of this is the early indication of sealevel rise at 19,000 yr BP (Fig. 9), which was probably caused by the loss of much of the Barents Sea ice dome (Fig. 30). Melting of the Barents ice dome at that time was inferred by Jones and

Keigwin in 1988 from negative $\delta^{18}O$ values of foraminifera in the northern part of the Nordic Sea. Denton and Hughes assign an equivalent maximum sealevel rise of 15 m to the Barents ice dome, which is consistent with the step up in sealevel shown in Figure 9 about 19,000 yr BP. However, this deglaciation was apparently not an essential part of the sequence of events causing the transition to interglacial oceanic circulation and climate.

The 6500 year interval of deglaciation that resulted in the final transition to strong conveyor-belt circulation was not one steady progression as the ice disappeared. Instead, the record from northern North Atlantic regions shows that it occurred in three contrasting phases, with an abrupt switch between each phase:

Phase 1 was an interval of temperature in the northern North Atlantic and in parts of Europe during Heinrich event H1 that was colder than earlier during the glacial maximum. This interval was known in Europe as the Older Dryas, after a widespread Arctic flower growing in Europe then. Nevertheless, despite the colder climate, the Fairbanks sealevel curve (Fig. 9) suggests that sealevel rose slowly over almost 3000 years, probably due to the zonal circulation and the associated dry climate deglaciation.

Phase 2 was an interval of warm sea surface in the northern North Atlantic with an early version of the warm conveyor-belt circulation and climate known in Europe as the Bølling-Alleröd period.

Phase 3 was another interval with a very cold northern sea surface that was associated with the cold Younger Dryas interval in Europe that ended with the final renewal of strong conveyor-belt circulation.

We now can explain the main features of this complex series of changes. To state the core of the argument first: The icebergs and meltwater from the southern edge of the Laurentide Ice Sheet kept the sea-surface temperature and salinity low in the North Atlantic in a zone adjacent to the warm Gulf Stream. This temperature contrast caused a west-to-east midlatitude jet stream that steered storms eastward and minimized rainfall and snowfall on the Nordic Sea and the northern gyre. This effect caused the salinity in the gyre to rise on two different occasions during the deglaciation, each time initiating strong lower-level deepwater formation.

In Phase 1, during the first thousand years of event H1 from about 17,500 to 16,500 yr BP, a veritable armada of icebergs was discharged from the central Laurentide Ice Sheet region into the northern gyre by way of Hudson Strait (Fig. 44). Most of these icebergs melted in midocean in the northern gyre. During the second part of H1, the icebergs came from Newfoundland and the coastal continental shelves to the southwest (Figs. 44 and 45). Most of these melted in midocean near the south edge of the gyre and the north edge of the Gulf Stream. Ice-rafted deposits of debris several centimeters thick that melted out of icebergs from both sources are found all across the Atlantic between latitudes 40° and 50°N. Smaller amounts are found in core NA87-22, northwest of Ireland, and core SU81-18 (Fig. 44), close to southern Portugal.

As mentioned before, these armadas of icebergs caused two effects that would have cooled the northern North Atlantic. When icebergs melt, their heat of fusion plus the specific heat of their meltwater would be about nine times

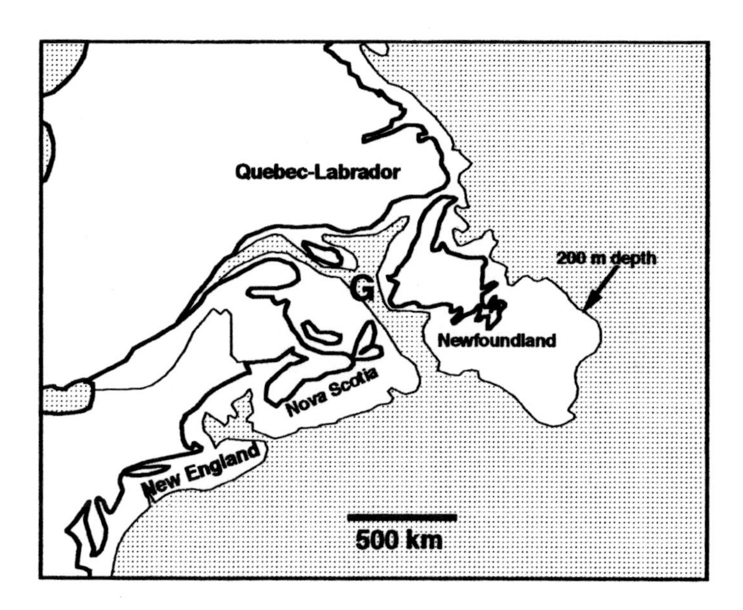

Figure 45: Largely exposed continental shelves at 19,000 yr BP at the last glacial maximum. The New England and Nova Scotia continental shelf areas were covered with thick ice sheets extending out from the higher inland elevations. The full extent of the ice on the Newfoundland shelf is not known. G is the Gulf of St. Lawrence.

as effective in cooling surrounding 15°C ocean as mixing in the same weight of water at 0°C. More importantly, their meltwater would have lowered the sea-surface salinity in the northern gyre and reduced the already weak conveyor-belt transport of warmer water that occurred at the time of the glacial maximum. These effects were amplified by a positive feedback factor discussed in Chapter 15. The cold, low-salinity surface water where the icebergs melted at this time was a barren area with very little sea life in the sea surface according to CLIMAP studies. We therefore infer that the massive parade of icebergs had shifted the atmospheric circulation pattern from a southwest-to-northeast flow over the North Atlantic, which had been favorable for growth and maintenance of the ice sheets, to the same flat, zonal, west-to-east flow that had caused the anomalous deglaciation 140,000 years ago. Once again, the lack of moisture and lack of cloud cover in ice-sheet regions provided the sunny summer days that were needed to tip the balance to deglaciation. In addition, just as in the anomalous deglaciation, the cold Atlantic north of the Gulf Stream provided the summer contrast with warm land surfaces in Europe that steered the jet stream to the northeast over western Europe, bringing warmer dry air to melt the Eurasian Ice Sheet.

Bard and colleagues in 2000 described the evidence for the H1 sequence of two parts: The discharge from Hudson Strait was identified in the ice-rafted debris of core SU81-18 off Portugal by the dominance of carbonate particles, which came from the limestone rocks on the floor of Hudson Bay and Hudson Strait. As the Hudson Strait icebergs diminished, the limestone particles were replaced by quartz and feldspar particles and some with red hematite coatings from southeastern Canada. These particles came from the deglaciation of Laurentide ice that had extended over the coastal continental shelves east of New England, Newfoundland, and the Maritime Provinces of Canada. The low sealevel at the last glacial maximum had exposed to the atmosphere the shelf areas from southern New England to the banks off Newfoundland, a distance of over 1500 km (Fig. 45). As the discharge of icebergs through Hudson Strait ended about 16,500 yr BP, the ice sheets on these continental shelves had become partly immersed by rising sealevel, and the buoyant increase in iceberg calving began to remove large amounts of ice from this southeast sector of the Laurentide Ice Sheet. The ice and meltwater from this sector maintained a mid-latitude zonal oceanic and atmospheric circulation and the associated dry-climate deglaciation. The glacial ice that covered Nova Scotia was particularly massive, where the continental shelf that is now beneath the sea is over 220 km wide (Fig. 45). The resulting icebergs and meltwater were the main cause for the long duration of event H1, and ice-rafted debris from these coastal rocks was also deposited on the sea bottom over the mid-Atlantic.

A decisive difference between the discharge of icebergs and meltwater from Hudson Strait and the discharge from the Maritimes was that the flow from Hudson Strait was largely confined to the northern gyre and looped to the north, while the discharge from the St. Lawrence area tended to remain at the edge of the Gulf Stream, and looped to the south. After the iceberg surge from Hudson Strait ceased, this division of flow kept meltwater out of the northern gyre, and enabled the evaporation losses and the low precipitation rates to increase the sea-surface salinity in the northern gyre during the second part of Phase 1.

In summary, the effect of the Heinrich discharge of icebergs from the heart of the Laurentide area through Hudson Strait was to initiate the deglacial process by decreasing moisture supplies to the ice sheets and by raising sealevel. This rise flooded and destabilized the mid-latitude ice sheets that had extended out on the continental shelf in New England, the Maritimes, and Newfoundland

(Fig. 45). The melting of this massive quantity of midlatitude ice then maintained the zonal oceanic and atmospheric flow over the North Atlantic leading to strong conveyor-belt oceanic circulation at the beginning of Phase 2 as the salinity in the northern gyre increased. However, note that while the Maritime coastal ice sheets were deglaciating, any slight increase in deepwater formation in the Nordic Sea drew lower-salinity replacement water northward from the midlatitude zone, and its low salinity prevented further increase in deepwater formation. Therefore, until the midlatitude discharge ended, the salinity in the Nordic Sea could rise to (but not above) the threshold for strong deepwater formation.

The midlatitude icebergs were injected into and melted in a zone at the edge of the Gulf Stream (Fig. 44) that was ideally suited to cool the sea and cause a strong temperature contrast across the edge of the Gulf Stream, thus maintaining the zonal storm tracks and eastward atmospheric flow and the aridity of the northern gyre. A back-of-the-envelope estimate of the cooling, taking into account the estimated rate of iceberg discharge, suggests that the coastal icebergs would have been more effective in providing a strong temperature contrast with the warm Gulf Stream than a larger amount of meltwater from the Hudson River Valley that would have mixed into the Gulf Stream about this time. The discharge of both the coastal ice (Stea and colleagues, 1996) and the Hudson Valley meltwater (Clark and colleagues, 2001) ceased at approximately 15,000 yr BP. The midlatitude water entering the gyre then no longer inhibited deepwater formation, the critical Nordic Sea salinity threshold was exceeded, and the strong conveyor-belt circulation began, thus initiating the warm Bølling-Alleröd interval of European climate about 14,800 yr BP (Fig. 46).

The triggering of the strong conveyor-belt flow caused a loss of zonality of atmospheric circulation as storm paths began to shift northward, and probably involved a race between the increasing precipitation on the gyre and Nordic Sea, which tended to lower the salinity, and the increasing northward transport of salt, which increased the gyre salinity as the conveyor-belt flow increased. But the race was won by the conveyor-belt in not more than about a half-century (Fig. 46). From the sea-surface temperatures indicated by warm and cold species of foraminifera, it appears that the oceanic polar front retreated first on the extreme eastern side of the Atlantic as the conveyor-belt flow increased. This southwest-northeast orientation of the front was due to the coriolis effect on the flow of replacement water, and it kept the polar-front jet stream and its storm paths well to the east, which minimized precipitation on the gyre and kept its salinity high during the transition to the strong conveyor-belt circulation.

In Phase 2, the warm Bølling-Alleröd interval lasted more than a thousand years, and with its warmth came a more rapid deglaciation. However, there were some notable brief fluctuations in the conveyor-belt strength caused, for example, by freshwater released in a jokulhlup from an ice-dammed lake in the Baltic Sea, discussed by Bodén et. al. in 1997. During the first few hundred years of Phase 2, all the Laurentide meltwater from the southern front of the ice sheet flowed into the Mississippi River and into the Gulf of Mexico, with no significant effect on high-latitude salinity. The southern edge of Laurentide ice slowly withdrew northward as deglaciation progressed, and about 14,000 yr BP, in the northeastern United States a drainage channel to the mid-Atlantic again opened up across northern New York and down the Hudson River valley. This added enough freshwater to the North Atlantic Drift to weaken the conveyor belt, as suggested by decreasing sea-surface salinity off Portugal, estimated by Duplessy et al. in 1992. This cooled the latter part of the Bølling-Alleröd interval somewhat, as shown by the $\delta^{18}O$ record of Greenland ice. But on the whole, northward transport of oceanic heat remained strong and the southern edge of the Laurentide Ice Sheet continued its northward retreat.

Phase 3 began at 13,500 yr BP (approximately) when the Hudson River channel started to be abandoned, and the Great Lakes meltwater began to drain toward the North Atlantic by way of the St. Lawrence River channel and the Gulf of St. Lawrence (Fig. 45). As the St. Lawrence channel opened up, the salinity off Portugal decreased and around 13,000 yr BP the salinity in the North Atlantic Drift fell far below that needed to enable the conveyor-belt to operate. Most of the conveyor-belt was shut down, and the slower mode of dry-climate deglaciation was restored. At this time about 12,700 yr BP, the 1200 year-period of intense cold in Europe known as the Younger Dryas began. The flow of meltwater from most of the southern front of the Laurentide Ice Sheet initiated this cold interval by re-establishing the zonal atmospheric circulation across the midlatitude North Atlantic. This again deprived the ice sheets of moisture and the deglaciation continued despite the colder temperatures. However, the southern Laurentide meltwater may have had

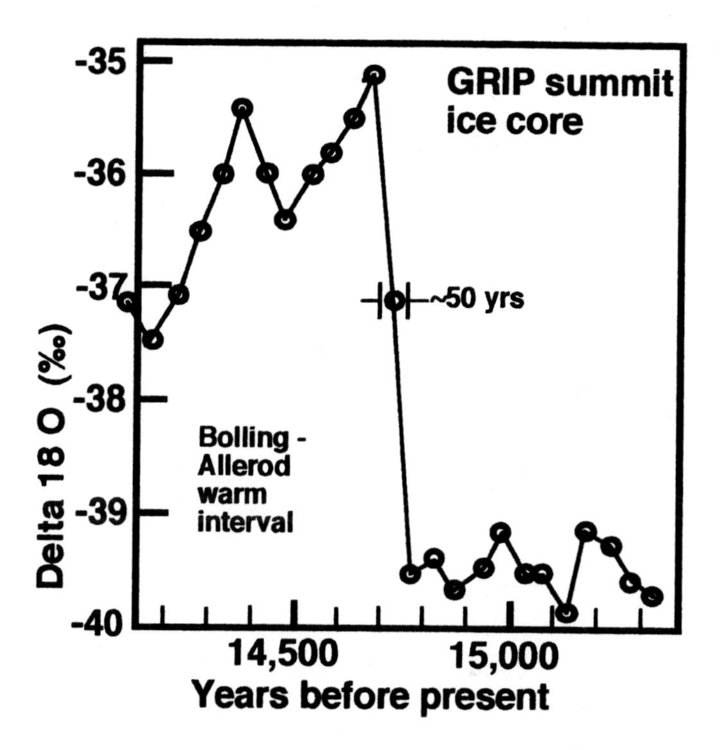

Figure 46: A warm climate switch in about 50 years on Greenland. Modified from Bond et al. (1993). The North Atlantic oceanic polar front moved northward as the conveyor-belt circulation accelerated in response to the initiation of strong lower level deepwater formation in the Nordic Sea. The storms that follow along the front moved rapidly closer to central Greenland, bringing snowfall from water vapor that had undergone less isotopic fractionation and was richer in ^{18}O. The data are from the GRIP core, but the transition age on the timescale is an average from the GRIP and GISP2 cores.

some help in maintaining the low temperatures of the northern North Atlantic, probably in the form of an iceberg discharge from the Arctic or from Hudson Strait, because sea temperatures remained very cold in the gyre west of

Ireland for a few hundred years. This extreme cold eventually diminished, and an analysis of populations of surface-water foraminifera show that the summer sea surface off Ireland became warmer than off Portugal, which remained under the influence of meltwater from the cold St. Lawrence outflow, with associated cloudy conditions.

As the end of the Younger Dryas (Phase 3) approached, the low salinity off Portugal and high salinity west of Ireland resembled conditions just before the conveyor-belt start-up at the beginning of the Bølling-Alleröd warm period 3000 years earlier. For a few hundred years before the end of Phase 3 at 11,500 yr BP, the flow of midlatitude meltwater through the Gulf of St. Lawrence consisted of all the meltwater from the southern front of the Laurentide Ice Sheet. With the retreat of the ice, less meltwater was coming from the eastern part of the ice-sheet front, and the Lake Agassiz outflow from the western sector probably contributed a major part of the total. This strong outflow through the St. Lawrence valley maintained the zonal circulation over the Atlantic and the dry-climate deglaciation, and allowed salinity of the northern gyre to rise again to the critical threshold.

The strong conveyor-belt flow was finally triggered by a re-diversion of Lake Agassiz meltwater into the Mackenzie and Mississippi Rivers (Fig. 47). A re-advance of Laurentide ice over Lake Superior at 11,500 yr BP and the slow rebound of the earth's crust after the general retreat of the ice sheet accomplished this. The much-reduced meltwater flow via the Gulf of St. Lawrence allowed the gyre salinity to exceed the critical threshold of

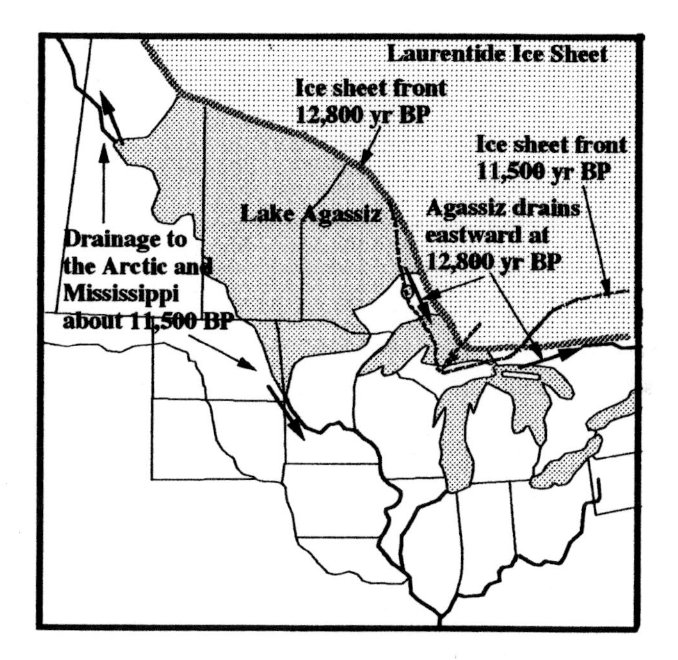

Figure 47: Diversion of Lake Agassiz to the Arctic and Mississippi River by re-advance of the Laurentide ice across Lake Superior. The drainage of Lake Agassiz toward the Gulf of St. Lawrence stopped about 11,500 years ago, thus allowing midlatitude sea-surface salinity to increase enough to trigger the North Atlantic interglacial mode of strong conveyor-belt deepwater formation in the Nordic Sea. Dates of the ice-sheet front are calendar years converted from carbon 14 years of Dyke and Prest (1986) using the results of Stuiver et al. (1998).

about 35.3‰ north of Ireland, which then triggered the strong conveyor-belt flow. As before, the Greenland ice core $\delta^{18}O$ indicates an abrupt transition in a few decades to the interglacial climate circulation. The large supply of excess salt carried northward maintained the flow to the present time, despite continuing smaller meltwater discharges during the latter part of the deglaciation.

The general climate warming brought about by the strong conveyor-belt circulation was quite effective in removing the Eurasian ice and ice from the southern Laurentide regions. But that was not the only significant factor that drove the deglaciation after 11,500 yr BP. Deglacial meltwater continued to enter the Labrador Sea by way of Hudson Strait with particularly large amounts after 9000 yr BP during the collapse of the Hudson Bay-James Bay ice dome. This freshwater would have prevented any deepwater formation in the Labrador Sea, keeping its surface cold. The polar front therefore lay farther to the southeast than today, and this minimized precipitation on the shrinking Laurentide Ice Sheet. Thus, a dry regional climate also contributed to the deglaciation.

How could this deglacial meltwater, which exited Hudson Strait from the very cold heart of the Laurentide Ice Sheet have been produced ? Melting of the ice-sheet surface that was exposed to air in this coldest part of the continent was probably not significant. The meltwater source was mainly the benthic melting of icebergs in Hudson Bay after calving from the ice dome. Melting within the frigid limits of the bay contrasts greatly with the situation earlier when Laurentide icebergs would float most of the distance across the Atlantic before finally melting away, and an explanation is needed.

The Hudson Bay ice dome would have accumulated while ice flowing off the Quebec-Labrador ice dome blocked the strait. As deglaciation progressed, the blocking ice that was standing in Hudson Strait in water a few hundred meters deep began to disintegrate because of the warmer gyre water and the buoyant forces caused by the rising sealevel. When the retreating cliff of ice reached the west end of the strait, the ice of the central dome standing in Hudson Bay also began to break up. However, the air temperatures of the central dome would still have been extremely low, and the critic's question would be: Why did not floating ice shelves develop along the edge of the dome and thus prevent the rapid collapse by buoyancy effects ? Such shelves certainly stabilized the cross-flowing ice at the mouth of Hudson Strait during normal glacial times when the northern gyre was quite cold. However, after the strong conveyor-belt circulation became re-established about 11,500 yr BP, the gyre water was much warmer. Therefore, although the low-salinity surface layer on the Labrador Sea was cold, the deeper seawater that came into contact with the underwater surface of the ice occupying Hudson Strait was able to supply a significant amount of heat to prevent formation of floating ice shelves and to melt the ice in the strait.

But farther back in the landlocked Hudson Bay the heat to melt the icebergs would have been more difficult to obtain. In fact, the collapse of the Hudson Bay ice dome was amazingly rapid, considering the confining geometry of the bay and the long distance from the source of the warmer seawater at the east end of the strait. The likely cause for this speedy collapse was a system of exchange currents in Hudson Strait. As the Hudson Bay dome disintegrated, the resulting meltwater formed a stratified low-salinity surface mixture overlying saltier water of greater density. Consequently, exchange currents would have developed analogous to the currents through the Strait of Gibraltar. There, the volume rates of flow of the inward and outward currents are quite large compared to their net difference, which is the hydrologic deficit of the Mediterranean. In the Hudson Bay case, the volume rates of flow would have been large compared to a difference in flow rates due only to the melting of dome ice that collapsed into the bay. Consequently, the large amount of slightly warmer seawater from the northern gyre, flowing in beneath the low-salinity layer at the surface, brought a steady supply of heat to enable the melting. Floating ice shelves could not form, and the icebergs from the dome melted within the bay. So, the collapse of the ice dome was not due to regional climate warming, but instead was mainly caused by attack of seawater that had been warmed by the restored conveyor-belt circulation and carried to the ice by the Hudson Strait exchange current.

After the larger part of the central ice dome had collapsed, significant meltwater discharges into the bay continued from remnant ice domes on Quebec, Keewatin, and Baffin Island until after 7000 yr BP. During that time, freshwater injection into the Labrador Sea would have continued to prevent deepwater formation there. Therefore the sea surface remained less saline and cold, with the oceanic polar front and storm paths farther to the southeast than today. Consequently, the final Laurentide deglaciation was brought about by a much-reduced moisture supply. Higher insolation was not the major factor.

It is now possible to suggest an answer to the question posed in Chapter 11: Why was there not a transition to strong conveyor-belt circulation at the end of the anomalous deglaciation? The likely answer is that, when the Mediterranean outflow was restored 6000 years after the anomalous deglaciation began, there was not an adequate amount of midlatitude meltwater being discharged to establish the zonal atmospheric flow. Therefore the critical threshold of high salinity in the gyre and the Nordic Sea was never attained. The St. Lawrence valley might have been blocked by ice, or the ice had retreated from the eastern sector of the Great Lakes with the western sector and Lake Agassiz still draining to the Mississippi.

One other question remains to be addressed: How large was the sea-surface temperature contrast in the midlatitude zone that kept the jet stream and traveling storms moving eastward over the mid-Atlantic during the Younger Dryas ? This question is important because a strong contrast is needed for the zonal atmospheric flow, which maintained deglaciation under more arid conditions in the ice-sheet areas. We can infer the temperature contrast with confidence. The meltwater from the entire southern front of the Laurentide Ice Sheet would have been rather cold when it encountered the cold Canadian Current off Newfoundland to form a stratified, low-salinity flow moving eastward, sandwiched between the northern gyre flow and the Gulf Stream. The low temperature of the eastward oceanic flow would have been reinforced by a smaller amount of northward transport of atmospheric heat due to zonal atmospheric flow. The temperature difference between the meltwater mixture and the adjacent Gulf Stream is indicated by temperatures derived from foraminiferal assemblages in core SU81-18 off southern Portugal reported by Duplessy et al. in 1992. Before the transition and with a midlatitude meltwater mixture flowing southward off Portugal, the summer temperature was 13°C. Immediately after the transition and with a strong North

Atlantic Drift, it was mostly Gulf Stream surface water like today that flowed southward off Portugal, and the foraminiferal species recovered from the sediment beneath indicate a summer temperature of 20°C. Therefore, across the entire North Atlantic up to the time of the transition a sharp 7°C contrast would have occurred over the northern edge of the Gulf Stream. This temperature step kept the jet stream and storms moving eastward along the oceanic front, and kept precipitation low on the northern gyre and Nordic Sea until the abrupt renewal of strong conveyor-belt flow.

It is now clear that the progression from the last glacial maximum about 19,000 yr BP to the warm interglacial oceanic circulation occurred by means of changes independent of the effects of Milankovitch insolation. Each of the transitions to the interglacial conveyor-belt circulation at about 14,800 and 11,500 yr BP occurred because of salinity increases in the northern gyre that were a consequence of zonal midlatitude oceanic and atmospheric flow caused by southern meltwater from the Laurentide Ice Sheet, probably with Hudson Strait blocked by a crossflow of ice from the Quebec-Labrador ice dome. These transitions did not require added solar energy inputs. And in the absence of the warm conveyor-belt, the southern Laurentide Ice Sheet bootstrapped its own deglaciation with meltwater that maintained zonal circulation.

The discharge of icebergs and meltwater through Hudson Strait, known as Heinrich event H1, initiated the deglacial sequence of events with no causal connection to high insolation. Was H1 a random event ? Why did the H1 event occur when it did ? The answers point toward a neo-Milankovitch hypothesis of orbital influence on Laurentide Ice Sheet growth and deglaciation.

14

Heinrich event dynamics

Six times during the last 120,000 years, massive quantities of icebergs drifted across the northern North Atlantic, usually from Hudson Strait. These icebergs cooled the northern gyre for a thousand years or more, and distributed thick layers of glacial debris on the seafloor as they melted. However, only Heinrich event H1 initiated the chain of deglacial events that triggered a transition to a warm interglacial climate. That deglaciation is known as termination 1. To grasp the importance of event H1 in bringing about our present warm climate, it is necessary to examine the cause of the event and others like it.

A cold Heinrich-like event also occurred just before the major deglaciation of termination 2 (Fig. 42). As mentioned earlier, this cold interval was reported from Denmark sediments by M. Seidenkrantz in 1993. Termination 2 is the conventional end of the Illinoian-Saalian ice age that was marked by a large rise in the sealevel at 128,000 yr BP and the climate warming of the last interglacial period. There isn't much detailed evidence of the sequence of changes during termination 2 at that distant time, but we can boldly assume that the Seidenkrantz event also triggered a deglaciation and stopped the falling sealevel at an elevation about 80 m below present, eventually resulting in the warm-climate transition.

What did these two events have in common ? Both occurred after rapid ice-sheet accumulation, H1 after about 10,000 years of ice-sheet growth with sealevel falling from

50 m to about 120 m below today, and the Seidenkrantz event after about 7000 years of sealevel fall from the +7.4 m level of the anomalous deglaciation to the -80 m sealevel of Aladdin's Cave, dated by Esat et al. to 130,000 yr BP. In each case, the low world-sealevel implies an extension of the Laurentide Ice Sheet far to the south.

Let us also assume that both events were caused by unstable conditions in the Laurentide Ice Sheet that caused its collapse, with an ice stream from the large central area rapidly surging into the Labrador Sea through Hudson Strait. The model proposed by MacAyeal in 1993 to explain this collapse begins with a thin ice sheet, frozen to the bedrock. As the thickness increases the bed becomes more insulated from the cold atmosphere, geothermal heat from the interior of the Earth melts the base layer of the ice, and sliding begins. After some time, glacial erosion accumulates sediment at the base of the ice, and with this lubrication the sliding instability becomes dominant and a long interval of surging occurs that drains out a large part of the central Laurentide ice. This is followed by the next cycle of buildup and collapse.

An alternative model of cyclic ice-dams across Hudson Strait can be developed from the discussion of Johnson and Lauritzen, who in 1995 examined the great jokulhlaup of Lake Zissaga. That jokulhlaup involved the collapse of a dynamic ice dam formed by ice flowing across Hudson Strait (Chapter 12). The alternative model includes the cyclic blockage of ice flowing from the Laurentide interior, and it seems to be quite consistent with the ice-core $\delta^{18}O$ data at several points. The crossflowing ice came from the Quebec-Labrador ice dome on the south (Fig. 1). It flowed northward through Ungava Bay, crossed the strait, and

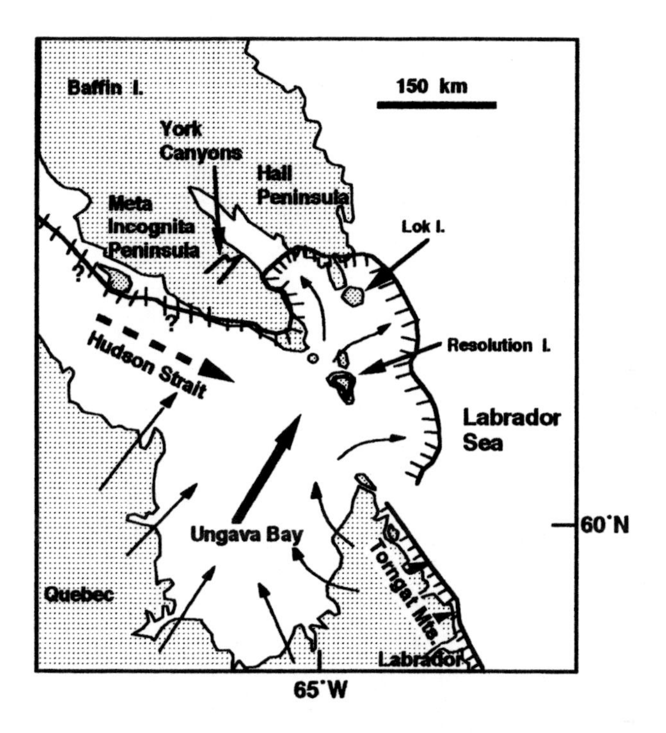

Figure 48: Hudson Strait ice dam, 9900 years BP. Modified from Stravers, Miller, and Kaufman (1992), with permission. The crossflow of ice on this occasion did not apparently reach as high as the York Canyon spillways, even though it extended to the north as far as the Hall Peninsula. Whenever the crossflow (heavy black arrow) was dominant during the last ice age, the axial eastward flow (heavy dashed arrow) from the Laurentide Ice Sheet interior was blocked.

piled up on the Meta Incognita Peninsula of Baffin Island (Fig. 48). In this cyclic mechanism, the crossflow of ice from the ice dome, where snowfall was heavy, competed with the eastward flow down the axis of the strait from the central Laurentide region, where the annual rate of snowfall was much less. When the crossflow was dominant, the eastward flow stopped and no ice left the central region.

When the central ice built up on the ice domes and in the strait to the west, it eventually overcame the crossflow, and rapid eastward axial flow occurred.

The crossflow discharged ice to the Labrador Sea at a relatively slow rate. The ultimate delivery of ice to the sea was by way of floating ice shelves, which grew onto the seaward fringe of the crossflow in the cold glacial climate, and were analogous to those in the Antarctic today. These shelves dynamically stabilized the dam by keeping the ice-sheet's surface slope small along the grounding line. After a few thousand years of accumulation, the central ice in the west end of the strait rose to a level above the crossflow and overpowered it. With the change from a slow crossflow to a rapid eastward flow, the stabilizing ice shelf at the mouth of the strait was broken away and, driven by the large excess accumulation in the central region, the interior ice surged into the Labrador Sea as the Heinrich event began. As the excess ice drained out of Hudson Strait, the surface elevation of the ice stream in the strait fell and the stream would have begun to float. The surge would have continued until the excess ice in the central Laurentide region had been removed, or until the eastward surge diminished and the crossflow could again dominate.

Earlier workers viewed the discharge of Laurentide ice through Hudson Strait as a steady flow down the axis of the strait from the west, like ice streams in the Antarctic today. However, there is little or no evidence for a massive grounded ice stream flowing eastward in the strait. There is no visible evidence for large-scale ice-sheet erosion on the Meta Incognita Peninsula that would indicate longterm flow parallel to the axis through the strait, nor are there significant glacial grooves in the bedrock in that direction that would indicate much recent parallel flow. Therefore, when the axial ice-stream surge did occur, the ice-stream border may not have reached the present shoreline, and the stream was likely floating in the eastern part of the strait with its surface not above present sealevel. This unpublished concept of competing cross and axial ice-flows was conceived after the publication of earlier ideas in a 1998 review by J.T. Andrews. The concept is an attractive working hypothesis, and is quite consistent with some of the detailed record of the isotope ratios from the ice cores on the Greenland ice cap.

The observed glacial erosion features and recent striations on rocks on the southern coast of Baffin Island are consistent with ice flow coming off the peninsula and also across the strait from Ungava Bay, and the past existence of such dams across the strait is not in doubt (Kaufman et al., 1993). A dam that broke at 116,000 yr BP and released a large amount of impounded water (Lake Zissaga) is inferred from the oxygen-isotope evidence and the coral-terrace record on Barbados as discussed in Chapter 12. There is abundant evidence for two brief intervals of crossflow during the last deglaciation in the early Holocene (Fig. 48), the first about 10,900 yr BP. Limestone rocks were plucked from the bottom of the strait, caught in the crossflowing ice, and deposited across the east end of the Meta Incognita Peninsula. There are also strand lines at 365 m above sealevel, formed at the edge of an impounded paleolake near the York Canyons, which were spillways for past ice dams (Fig. 40). The spillways have been eroded into hard crystalline rock, and probably carried overflow water whenever ice flow from the south crossed the strait during the more active phases of each ice age over the last million or more years. During the occasions later in the last glaciation, of course, more glacial ice than water was held back by the ice dam.

This is probably why the ice dam occurred. The crossflow became strong because of the topographic constraints and the high rate of snowfall on the Quebec-Labrador ice dome, with surging as a possible independent process. The dome area is located near the Labrador Sea (Fig. 1) and in or close to the main paths of storms moving northeastward across the continent and along the coast. The precipitation rates on the ice dome would therefore have been much greater than in the interior centers of accumulation, Hudson Bay-James Bay, Keewatin, and the Foxe Basin-Baffin Island domes. The rapidly accumulating ice on the Quebec-Labrador dome was unable to flow eastward because of the Torngat Mountain range on the northern Labrador coast, and the plateau elevations in southern Quebec inhibited flow toward the south. Consequently, the main discharge path from the dome was through Ungava Bay and across Hudson Strait.

The crossflow early in the Holocene may have been a surge because, although it was so strong that it continued on across the tip of Meta Incognita Peninsula to Frobisher Bay and beyond to the Hall Peninsula and Lok Island on the north, it nevertheless was not a lasting flow. On other occasions, however, the crossflow was probably steady and sustained for long intervals. The crossflow was partly blocked by Resolution Island and the Peninsula (Fig. 48) thus causing the thickness to increase in the strait until the flow became grounded everywhere, even in a 900 m-deep half-graben pit in the strait outside of Ungava Bay. The pit today has very little till in it, having been cleaned out by the most recent crossflow. Whenever the ice dam was firmly grounded, neither ice nor meltwater could drain from the interior eastward through the strait. The surface of the dam could have been 350 m or more above present sealevel (neglecting minor isostatic effects) and the dam occupied an east-west area almost 200 km-long in the strait. Some of the crossflow accumulated on the west side of the dam. The remainder of the crossflow eventually entered the Labrador Sea by extension into floating ice shelves from which tabular icebergs occasionally broke away. The top of the dam at its maximum was above the elevations of the two York Canyon spillways (Fig. 48) because the York Canyons were eroded beneath glacial ice. With the dam in place the spillways were the only drainage outlets for meltwater accumulating to the west. Because of the dam, excess glacial ice would have accumulated on all the Laurentide ice domes to an added thickness much greater than 350 m, and over a very large area.

Although the accumulation of interior ice always cyclically overpowered the crossflow, the crossflow was probably weakened by reduced precipitation before the change to axial flow occurred. As the Laurentide ice sheet extended farther to the south and as its elevation increased, storms steered by land-ice-sheet temperature contrasts would have tended to pass more to the south of the Quebec-Labrador ice dome, and snowfall on the dome would have decreased. At some point the diminished ice flow into the dam became inadequate to prevent the accumulating ice on the west from overpowering the crossflow and initiating a long surging Heinrich event.

When ice rather than water was impounded west of the crossflow, the collapse of the ice dam would have been slow. Decades or even hundreds of years might have been needed to make the change to strong axial flow, because of the nearly 200 km-width of the dam, as measured along the axis of the strait. In time, however, the stabilizing ice shelves would have been swept away, the flow would have begun to surge, and the Labrador Sea and parts of the northern gyre would have been flooded with icebergs as the Heinrich event began. The gyre, which had been slowly cooling because of the southward diversion of storm tracks caused by the increasing height and extent of the ice dome to the south, became abruptly colder due to iceberg melting and the further reduction of the already limited production of deepwater. When the long surge ended and the crossflow again dominated, the elevations of the main ice domes in Figure 1 would have been much reduced. The salinity of the northern gyre then rapidly increased, restoring normal shallow deepwater formation and thus attracting storm paths back to the north and ensuring precipitation and a good crossflow for a considerable time.

Support for this proposed cycle is inferred from the $\delta^{18}O$ of Greenland ice cores (Fig. 49) in the middle of the last glaciation when the high latitude North Atlantic was still somewhat warm, and larger temperature changes were possible in the sea near Greenland and on the Greenland ice cap. The small upward spikes in the $\delta^{18}O$ at X, shortly before the isotope-ratio drop suggesting ice-dam failure, are consistent with a decrease in discharge of cross-flowing ice to the Labrador Sea as the ice dam became weaker, just before it allowed the switch to axial flow to occur. The slightly increased salinity in the gyre and the slight increase in conveyor-belt flow that would have resulted from a smaller discharge of icebergs from the crossflow probably

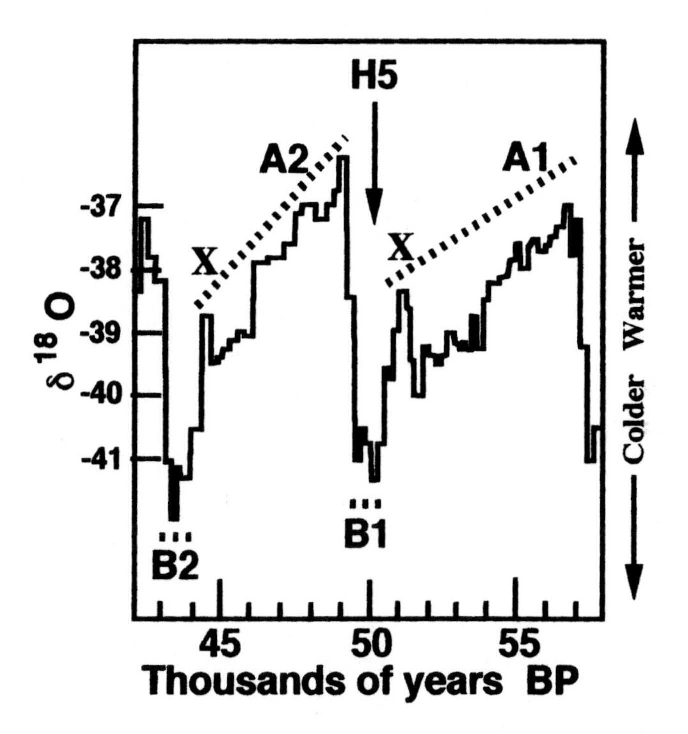

Figure 49: Indications of cyclic ice flows in Hudson Strait in the model of competing axial and crossflowing ice streams. Oxygen-isotope ratio variations in the GRIP Greenland ice core are a proxy for the latitudinal position of storm tracks south of Greenland. The farther south the storm tracks are, the more negative the ratios. The data are interpreted as a competition between the crossflow from the Quebec-Labrador ice dome and the axial flow eastward in the strait from the Laurentide interior, with resulting effects on the salinity and temperature of the Labrador Sea. See text. A1 and A2 - crossflow occurs, ice dam in place and Laurentide ice accumulates. B1 and B2 - axial flow occurs and a surging ice stream drains the central Laurentide ice mass. X marks a decrease in crossflow prior to the dam failure. H5 is Heinrich event #5. Modified from Bond et al. (1993).

caused this upward and less negative $\delta^{18}O$ spike in the Greenland ice core. Amongst the Heinrich events, there are shorter (2000-3000 years) Dansgaard-Öeschger cycles in the $\delta^{18}O$ of ice cores. These also are associated with ice-rafting events as Bond and Lotti (1995) showed by correlating peaks in deposits of rock particles from widely different sources with ice core $\delta^{18}O$ minima. The synchronous deposition from many sources could be caused by release of marine-based ice sheets along the coasts due to rapid sealevel rise caused by failure of smaller ice dams in Hudson Strait or in the Baltic Sea area.

In summary, Heinrich cycles suggest a mechanism that favors Laurentide ice buildup when salinity is higher in the Labrador Sea and subpolar waters to the east. Deepwater may form off Labrador, bringing saline replacement water into that area. There are good reasons to infer an unusually high salinity during the 4000 years before Heinrich event H1. Subtropical summer insolation was unusually low. African monsoons were quite weak, and the Mediterranean outflow would have been strong. A northward storm-track shift over ice-free sea surfaces under cold-climate conditions is suggested by a rise of over 2‰ in the baseline $\delta^{18}O$ of the GISP2 ice core (Bond et al., 1997) prior to 18,000 yr BP. More northerly-oriented storm paths are also consistent with the rapid growth of the Barents Sea ice dome that began about 22,000 yr BP, as shown by evidence in sediment from the Fram Strait reported by Hebbeln and colleagues in 1994. These storm paths would have maintained a healthy crossflow of ice across Hudson Strait, which prevented any outflow of accumulating ice from the central regions of the Laurentide Ice Sheet for a long time. The resulting long-lived ice dam across Hudson Strait therefore enabled extension of Laurentide ice far to the south before deglaciation was initiated by Heinrich event H1 at the maximum of the last glaciation.

15

A neo-Milankovitch hypothesis

There is no way that a Milankovitch insolation difference could have forced the major deglaciation from 144,000 to 137,000 years ago, when sealevel rose up to +7.4 m. And in Chapter 11 we saw that a major reglaciation occurred in violation of Milankovitch during a large increase in summer insolation from 137,000 to 130,000 years ago. Furthermore, in Chapter 13 it was argued that the entire deglaciation from 18,000 to 7000 yr BP consisted of intervals of dry climate alternating with strong conveyor-belt heat transport. None of these deglacial steps required high insolation. Yet over long intervals of ice-age climate, correlations of insolation with the $\delta^{18}O$ proxy for ice volume still imply a connection between orbital changes and ice-volume variation. The inference is that this occurs by way of the African monsoons and the Mediterranean Sea. There is also a previously unrecognized feedback loop in the climate system that can amplify the effects that insolation change can have on North Atlantic circulation. Now, we will combine these ideas to suggest the long-sought physical connection between insolation change and ice-age climate.

The anomalous deglaciation was associated with a zonal circulation in the North Atlantic, and it demonstrates that major deglaciation can occur by reducing the moisture supply to the ice sheets. Thus, we see that there are two modes of North Atlantic oceanic and atmospheric circulation in which major deglaciations can take place:

(1) The zonal mode with a cold northern gyre and little deepwater formation. Moisture is withheld from northern ice sheets by eastward movement of storms, guided by a zonal oceanic front in the midlatitude North Atlantic.

(2) The conventional interglacial mode with a warm northern gyre and strong deepwater formation. Northward movement of storms occurs with corresponding transport of much oceanic and atmospheric heat.

Between these two extremes of storm paths lies a path over the northern North Atlantic that is optimal for rapid ice-sheet growth. This path is slightly less northward-oriented than the modern jet stream track shown in Figure 50, and it supplies much moisture to both the Laurentide and the Eurasian Ice Sheets. Between the optimal path and the zonal extreme lie paths associated with intermediate salinities and temperatures in the northern gyre, with ice-sheet accumulation rates decreasing and finally becoming negative as the paths shift southward to the zonal extreme. The idea is that during the glacial intervals of ice-age climate, variations in the saline Mediterranean outflow move the jet stream and storm-track paths northward or southward within this envelope of possibilities causing the ice sheets to grow or shrink.

To show how this proposed mechanism works, we first examine African monsoons and their association with insolation variations. The feedback loop that amplifies monsoon effects is then described. Finally, some examples of this neo-Milankovitch effect of insolation on ice volume are cited, and a neo-Milankovitch version of the history of ice ages over the last three million years is outlined.

Once again, here is the fundamental chain of cause and effect that links monsoons to ice-volume variations:

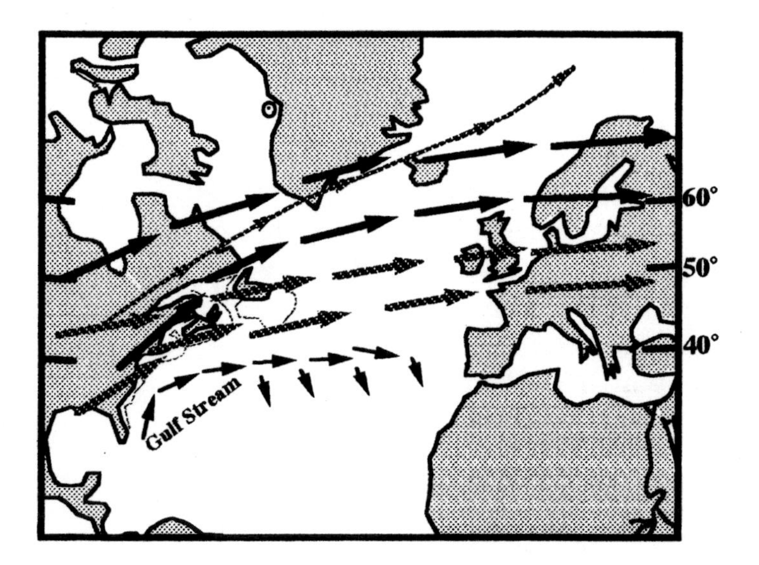

Figure 50: Winter storm-path differences due to monsoon effects during the ice age at times of African monsoon extremes. Large black arrows - weak monsoons, more salinity and warmth in the high latitude sea surface, optimal high rate of ice-sheet growth. Large hatched arrows - strong monsoons, less salty Mediterranean, less salinity in the high latitudes, Labrador Sea becomes colder with little deepwater formation, atmospheric flow is more zonally-directed, ice sheets receive less moisture and tend to grow more slowly or diminish. Small hatched arrows - modern storm paths in January.

Stronger monsoons cause larger Nile River flow to the Mediterranean and higher humidity in that region. As a result, salinity and outflow decrease, causing a decrease in high latitude sea-surface salinity. The moderate conveyor-belt of glacial times weakens, cooling the northern gyre. The average path of storms shifts southward, following the sea-surface temperature contrasts. This deprives the ice sheets of moisture, thus causing slower ice-sheet growth or deglaciation. To get a more precise idea of the monsoon correlation with insolation, we next examine some extreme examples of monsoon variations, as indicated by the extension of rains and moist climate to the north across the African continent.

Here are two examples of moist conditions: (1) When a level of 50 langleys per day greater than present at latitude 25°N occurred about 128,000 yr BP, sapropel #5 was forming. Monsoons extended into central Libya at 27°N and fed the Shati paleolake, which had a volume of 20 km^3 and covered 2000 km^2. (2) When the insolation was 40 langleys per day greater at 10,000 yr BP, sapropel #1 was forming. Monsoons extended north to Aswan in central Egypt at 24°N (Fig. 51), as reported by Yan and Petit-Maire. The Sahara was then an inhabited grassland, and 31 of 32 lakes scattered between the equator and the Mediterranean coast had lake levels either higher or much higher than today, as summarized by Street-Perrott and Perrott in 1993. Evidence of erosion in the upper Nile and an additional major Nile channel suggest an average volume of flow perhaps three times larger than occurs today. At such times, the weaker Mediterranean outflow implies little intermediate-level deepwater formation off Labrador, and the Labrador Sea would have been cold and icy.

In contrast, when insolation was at a minimum near the last glacial maximum (Fig. 51), the African monsoons extended only to the headwaters of the Nile at about 10°N, North Africa was dry, and the climate of the Mediterranean and Black Sea regions was quite arid. Although Mediterranean Sea temperatures were cooler than today, in the arid climate it is likely that evaporation losses were also large, as suggested by the low sealevel and reduced area of

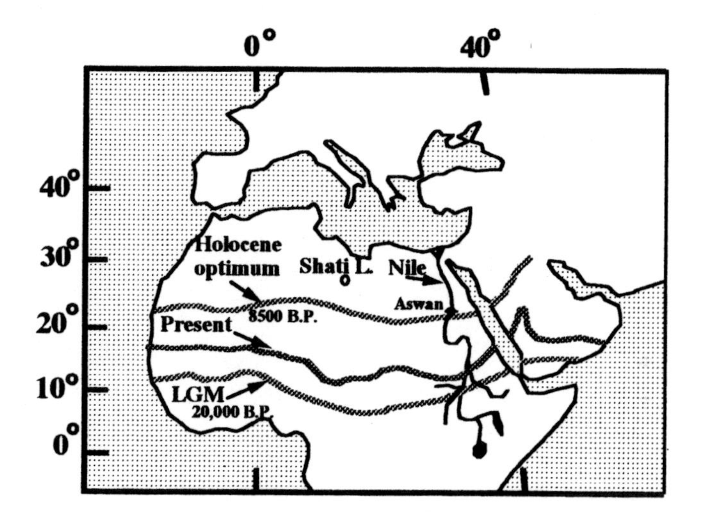

Figure 51: Northern extent of African monsoons. Modified from Yan and Petit-Maire (1994) with permission from Elsevier Science. The weak monsoons at 20,000 yr BP favored a salty Mediterranean Sea and ice-sheet growth. Likewise, the present day monsoons are also weak, and this is consistent with a threshold for new ice-sheet growth.

the isolated Black Sea. In addition, the low world sealevel at the glacial maximum diminished the cross section of the Strait of Gibraltar. The salinity of the outflow could therefore have been somewhat higher than today with a more effective delivery of excess salt to the high latitudes as suggested by a dashed arrow in Figure 29. The resulting higher salinity in the northern North Atlantic increased intermediate-level deepwater formation, causing more ice-free sea area and the rapid ice-sheet growth that preceded the last deglaciation.

Intuitively, the variations in monsoons seem quite large to have been caused by only a few percent change in the

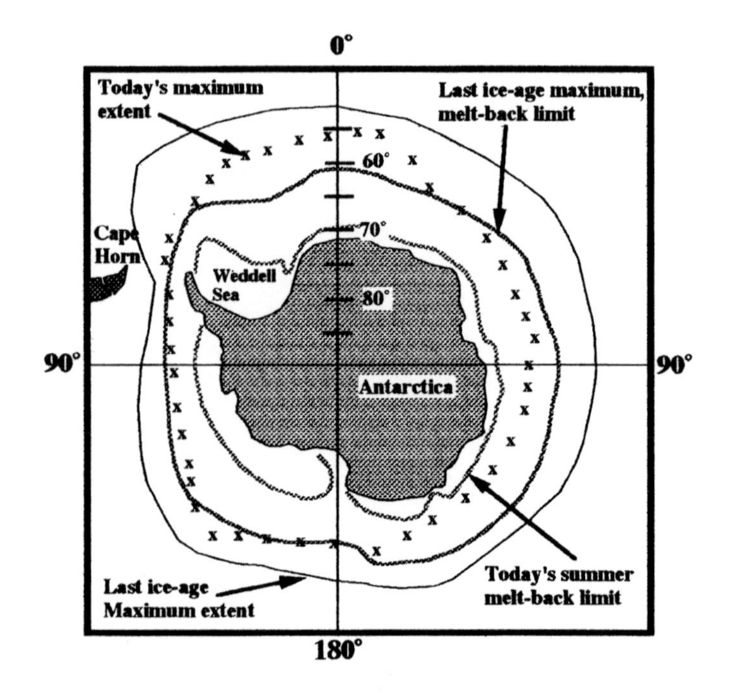

Figure 52: Extreme limits of sea-ice coverage around Antarctica. The last ice-age maximum extent is that depicted by the CLIMAP study of 1976, and the last ice-age maximum meltback limit is placed consistent with deep-sea sediment studies in the Atlantic and Indian Ocean sectors indicating that perennial heavy summer ice extended over 1000 km farther north than today.

insolation values of the Sahara desert. That cause becomes more convincing when one considers a proposed positive feedback loop that can amplify the insolation effects. The variable extent of sea-ice coverage in the Southern Ocean (Fig. 52), as discussed in Chapter 9, is an essential element in the loop. In modern interglacial times when the conveyor-belt is strong and the volume of salty and potentially more-dense water exported to the Southern

Ocean is large, the sea ice melts almost all the way back to the Antarctic continent in the summer in most areas. The relatively greater density of the Antarctic Water enables it to sink and flow away beneath the subtropical water at its northern boundary and this limits the accumulation and the thickness of the surface layer. At the ice-age extreme without the saline water from the north, the Antarctic Water is less dense and its boundary must have moved northward. The boundary did so, as discussed by Armand in 2000, and the thick summer sea ice persisted over a thousand kilometers farther north than today according to the CLIMAP 1976 report. The expansion would have been a major cause of cooling the Southern Hemisphere climate.

This southern climate variation amplifies the effects of insolation variations. To see how this occurs, it is necessary to consider the forces that drive the African summer monsoons. The fundamental driving agency is the temperature difference between the arid areas of North Africa and the Arabian Peninsula and the cooler winter areas across the equator to the south. The heated air over the warm land masses rises and must be replaced. The replacement flow comes from the land and ocean areas south of the equator that are cooled by the low winter insolation and low oceanic temperatures closer to the Antarctic continent. The cooler these areas are relative to the Sahara, the stronger are the monsoons.

The feedback factor that amplifies an increase in the temperature difference between the Sahara and Southern Hemisphere latitudes is therefore the expansion of the zone of Antarctic sea ice. The southern climate cooling that follows increases the latitudinal temperature difference across the equator that drives monsoons. Stronger monsoons reduce the Mediterranean salinity and outflow, which in turn reduces northern deepwater formation and saline water input to the Antarctic. This further expands the sea-ice zone, as depicted in the cartoon of Figure 53.

The feedback loop would have a characteristic delayed response to a step change, with probably as much as a few centuries needed to achieve a new steady-state condition in the loop. Consequently, the effects of durable events such as orbital insolation changes or Heinrich events would be amplified significantly, whereas brief events, such as minor glacial meltwater discharges may not be amplified very much. The cooling effect on the northern gyre caused by a Heinrich iceberg discharge is amplified around the loop in the same way as the effects of warmer northern summers in the Sahara. The initial effect on the temperature of the northern gyre due to the iceberg discharge in the first few decades may not be large. But after a few hundred years of amplification with decreasing Mediterranean outflow, the conveyor-belt may be nearly shut down, with the polar front near northern Spain, as it was during event H1.

Making use of A. Berger's tabulations, we can infer how orbital effects can change temperature differences across the equator that drive the monsoons. In simplified form, the two main orbital factors are involved: (1) The precessional heating or cooling of North Africa. (2) The effect of axis tilt on the temperatures of the more southerly Southern Hemisphere zones. When northern summers occur close to the sun, the Sahara at 25°N latitude and the Southern Hemisphere are both receiving increased insolation. The temperature increase is probably greater in the Sahara with the sun high overhead than it is to the south of the equator where the sun is much lower in

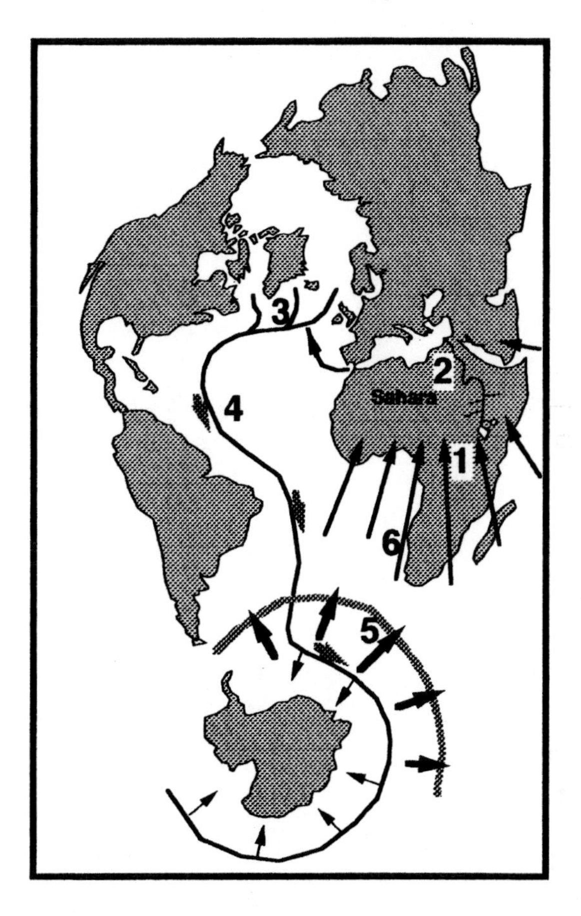

Figure 53: An amplifier for the effect of orbital insolation changes during glacial times. (1) Monsoon strength increases when summers occur close to the sun and the Sahara is warmer. (2) Nile flow increases, Mediterranean salinity and outflow decrease. (3) Intermediate-level saline deepwater formation decreases. (4) Excess salt in deepwater exported to Antarctica diminishes. (5) Less excess salt upwells around Antarctica, sea ice thickens and expands. (6) Southern climate cools, temperature difference driving the monsoons increases. (1) Monsoons become even stronger and amplify the initial decrease in Med salinity and outflow. Ultimate result: sapropels form and a more zonal North Atlantic circulation causes a minor deglaciation.

the sky. This alone would enhance the monsoons, but the temperature difference and the monsoon effect can also be increased or decreased by the tilt factor.

The polar-axis tilt effect (see Appendix) is strong in the south with the sun lower in the winter sky, but is weak near the equator. Therefore, one would expect that temperature differences from the Sahara across the equator during northern summer would be significantly influenced by tilt, and that is how it is. There is no easy way to quantitatively estimate the importance of these two orbital effects. Nevertheless, we can make use of Berger's tabulated insolation data and compare specific glacial trends with insolation difference profiles over a latitude range across the equator. This comparison enables us to test for consistency with the geological records when the precession factor is reinforced or diminished by the tilt factor. The most intriguing example in the records is the ice sheet buildup before the anomalous deglaciation that began about 144,000 yr BP.

During the last 250,000 years, the only 25°N insolation maximum for which there is no corresponding Mediterranean sapropel occurred during an interval around 150,000 yr BP (Fig. 42) when the precession factor greatly increased northern insolation while the axis tilt factor reduced it to some extent. Because the tilt was less than today for northern summers, the insolation in the polar latitudes north of 65°N was somewhat less. But south of latitude 65°N the strong precession effect increased summer insolation (Fig. 54). With 1950 AD as a zero-point reference, the insolation changes multiplied by the areas of the earth's surface at each latitude zone can be

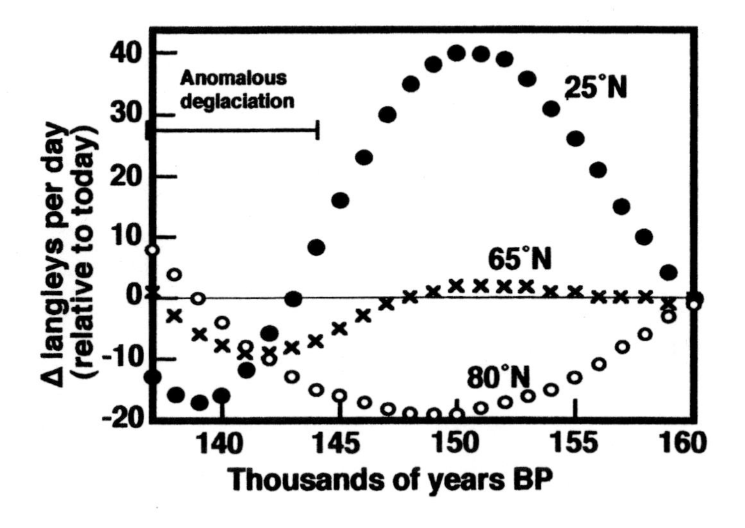

Figure 54: Trends of insolation around 150,000 yr BP for low and high latitudes before the anomalous deglaciation that began about 144,000 yr BP.

compared for the total areas of each of two latitude zones using Berger's tabulated insolation values. The result is that the zone to the south between 65°N and 30°N at 150,000 yr BP received eleven times as much added solar energy in the summer as was subtracted from the zone north of 65°N. During this time the Illinoian-Saalian ice age was approaching a maximum and most of the ice-sheet area was located south of 65°N. With the higher insolation south of 65°N, the expectation of Milankovitch would have been deglaciation, not ice-sheet growth. But the positive oxygen-isotope ratios of the sediments (Fig. 8) and absence of sapropel formation are consistent with the ice-sheet growth that led to the anomalous deglaciation that occurred from 144,000 to 137,000 yr BP.

This violation of Milankovitch is understandable in the neo-Milankovitch model in which the tilt influences the monsoons. The low-latitude Sahara insolation would have been only slightly affected by the opposed tilt, but simultaneously, the tilt caused Southern Hemisphere latitudes to receive more added winter insolation (Fig. 55) relative to today than was added at 25°N. Relative to the Sahara, this would have made the temperature differences over the latitude span across the equator even less than today, and so would have caused even weaker monsoons. This increased the Mediterranean outflow. The higher salinity in the gyre and resulting northward position of the jet stream and storm paths, as suggested in Figure 50, would have increased precipitation on the northern ice sheets for thousands of years. The stronger saline flow of intermediate-level deepwater to the Southern Ocean would have reduced the sea-ice coverage, thus reinforcing the effect of axis tilt on warming of southern winters. These conditions led to the ice blockage of Siberian rivers and the washout of the Bosporus sediment dam at 144,000 yr BP.

In the examples of 86,000 and 174,000 yr BP (Fig. 55), both precession and tilt increased the insolation during northern summer, but tilt decreased the insolation during the simultaneous southern winter. Consequently, the added warmth in the Sahara combined with less warmth in the southern latitudes to generate strong monsoons, a weaker outflow from the Mediterranean, and therefore more-zonal storm paths in a colder northern North Atlantic. This caused deglacial conditions in the Northern

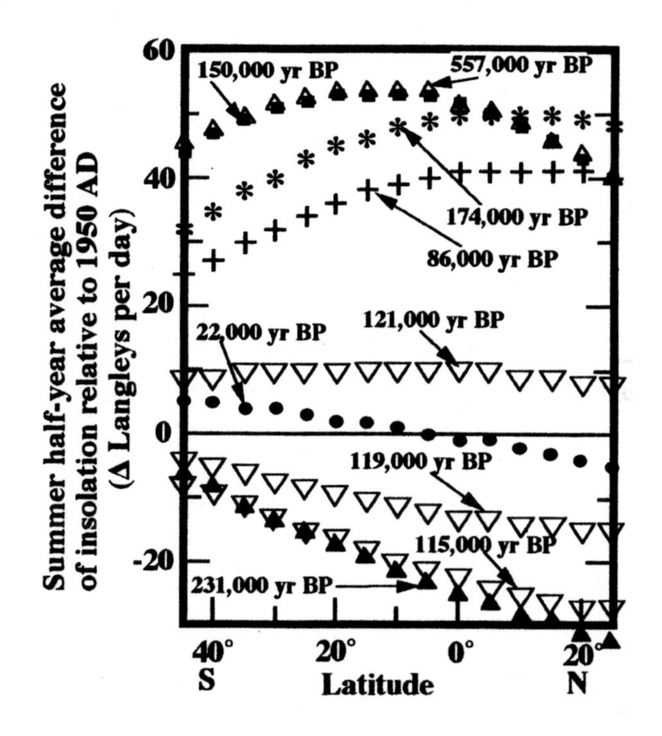

Figure 55: Axis tilt effects on insolation differences during glacial times. Caloric half-year insolation differences referenced to 1950 AD vs. latitude for 86,000 and 174,000 yr BP (tilt effect subtracts from precessional warming in southern winter) and 150,000, 557,000, 22,000, 119,000, 115,000 yr BP and 231,000 yr BP (tilt effect adds to precessional warming in southern winter.) Compared with today, southern hemisphere additions larger than those at 25°N imply relatively greater southern warmth, which inhibited monsoons, and favored higher Med salinity and storm paths of Figure 50 that were more optimal for glaciation. Conversely, lower southern additions at 86,000 and 174,000 yr BP imply relatively cooler surfaces in the south and stronger monsoons with lower Med salinity, thus causing more-zonal storm paths and moderate deglaciation. Heavy ice-sheet growth at 150,000 yr BP is contrary to Milankovitch but is consistent with weaker monsoons and a less-zonal circulation. See text. Data points from Berger's 1978 tabulations.

Hemisphere. Sealevel rose to about -14 m at 82,000 yr BP, but no transition to interglacial conveyor-belt circulation occurred because the Laurentide Ice Sheet was not extended sufficiently far to the south, as discussed earlier.

The three sets of large open triangles (Fig. 55) imply the shift toward warmer southern winters, relative to the Sahara, due to favorable tilt at the beginning of the last ice age. The trend was toward weaker monsoons, causing greater salinity in the Mediterranean and northern gyre, and a moderate but sustained conveyor-belt circulation with warmth in the gyre during the early glacial accumulation. The quite similar latitudinal differences at 70,000 yr BP (not shown) and 231,000 yr BP were likewise accompanied by sustained warmth in the northern gyre. The resulting optimum storm paths across North America, the North Atlantic, and Europe caused large volumes of glacial ice to rapidly accumulate, as discussed in great detail by W.F. Ruddiman and various colleagues in 1979, 1980, 1981, and 1982. These intervals of gyre warmth during ice-sheet growth are unexplained in the classical Milankovitch model of climate change, but occur naturally as a consequence of monsoon changes and the effect of axis tilt in the Southern Hemisphere in the neo-Milankovitch scheme.

These examples lead to the conclusion that effects of insolation changes caused by axis tilt and orbital precession can be amplified through the monsoons and the Mediterranean Sea salinity to modulate high-latitude salinity and oceanic circulation during glacial times in the absence of strong conveyor-belt circulation in the North Atlantic. This is the mechanism that produced the dominant periodicities of 41,000 and 23,000 years that were found in the oxygen-isotope proxy for glacial-ice volume in the classic study of Hays, Imbrie, and Shackleton in 1976. The effects of these orbital configurations may at times be overprinted by meltwater events, and consistency is not always obtained. Abundant meltwater effects are implied in the records of the GRIP ice core from central Greenland and from northern Northern Atlantic sediments, discussed by Bond et al. in 1993. Nevertheless, before the well documented sealevel maximum about 82,000 yr BP, for example, the records suggest several thousand years of relatively low sea-surface temperatures consistent with a more-zonal deglacial-type of circulation, as in Figure 50.

Note that during glacial times when stronger precession and stronger tilt effects occurred together in the Northern Hemisphere, both effects promoted deglaciation, but not because of a warmer climate, as in the Milankovitch hypothesis. Instead, we have another paradox. With the monsoon mechanism, the higher insolation at 25°N combines with a weaker tilt effect in the Southern Hemisphere, causing a more-zonal North Atlantic circulation, more aridity on the ice sheets, and cooler high latitudes. Higher insolation did not cause a net warming.

The effectiveness of the Antarctic sea-ice extent in cooling southern climate and consequently driving the monsoons can be inferred from the ending of major deglaciation in the area surrounding Hudson Bay at about 7000 yr BP. The resulting cessation of large meltwater discharges via Hudson Strait would have enabled the renewal of deepwater formation in the Labrador Sea, with an increase in saline deepwater exported to the Southern Ocean and a reduction in extent of sea ice there. This appears to have significantly reduced the monsoon activity, because the deposition of sapropel #1 also ended in the eastern Mediterranean about 7000 yr BP with the corresponding feedback by way of the high-latitude North Atlantic that would have increased deepwater formation and salt export to the Southern Ocean. A similar inference can be made from the re-appearance of foraminifera at the TR172-22 core site in the eastern Mediterranean at approximately 127,000 yr BP (Fig. 34), which implies a diminishing monsoon activity and reduced Nile flow. This may have been caused by the renewal of stronger conveyor-belt flow at that time, as suggested by the appearance of the pollen of warmth-loving trees in Germany shortly after 127,000 yr BP (Fig. 38). On both of these occasions the latitudinal profile of summer insolation from 25°N to 45°S was changing slowly and remained favorable for strong monsoons. It therefore appears that the insolation effects in the feedback loop were modified by the reduction of Antarctic sea-ice cover.

Finally, a persistent critic might ask, why is it that the warm-climate deglaciation occurred near the peak of strong northern insolation during the last deglaciation and also during the insolation peak following the earlier Illinoian ice age if the transition is not really dependent on strong Milankovitch insolation ? The answer lies in the large southward extension of Laurentide ice, and its cause. The effect of the somewhat warmer and more ice-free Labrador Sea on the dynamics of the Heinrich event H1 was discussed in Chapter 14. In this concept, the iceberg jokulhlaup of event H1 that occurred at the end of the last glaciation was the result of a large buildup of the central Laurentide Ice Sheet west of the ice dam in Hudson Strait. At that time low-latitude northern summer insolation was weak, with weak monsoons. The African aridity was amplified by the feedback loop, increasing Mediterranean Sea salinity and outflow, and thus causing a more saline and warmer Labrador Sea, northerly storm paths, heavy precipitation, and a strong flow of ice across Hudson Strait. After the deep insolation minimum, the cross-equatorial insolation gradient increased, causing the monsoons to strengthen. The Labrador Sea salinity began to decline, storm paths shifted, less snow began to fall on the Quebec-Labrador ice dome, and the crossflow weakened. Therefore, the breakup of the ice dam and the switch to surging ice flow eastward occurred when insolation was rising, thus creating the illusion of a classic Milankovitch deglaciation that began because of climate warming.

But it is indeed an illusion, because the deglaciation and switch to warm interglacial oceanic circulation was the result of the chain of events initiated by quite low Milankovitch summer insolation in the north and relatively higher tilt insolation in the south, giving the latitudinal profile of insolation differences at 22,000 yr BP shown in Figure 55, which caused rapid and extensive Laurentide Ice Sheet growth. Similarly, the switch to interglacial oceanic circulation about 127,000 years ago was the result of extensive ice-sheet growth that began at 136,700 yr BP (Fig. 16) under low 25°N insolation, and resulted in a Heinrich-type event about 130,000 yr BP. It was only because the higher Mediterranean salinity occurred beginning with a large nucleus of the Laurentide Ice Sheet in these examples that the ice became sufficiently extended to the south to provide the midlatitude meltwater and zonal atmospheric circulation that made possible the transition to the interglacial conveyor-belt oceanic flow.

So, to answer the critic, it is likely that over the last

million years, the major warm-climate deglaciations were completed most often when insolation was high because of the long delay associated with a progressive Laurentide deglaciation started by a key Heinrich-type event, after which it was necessary to melt an unusually large amount of ice, partly under dry-climate conditions. The timing depended on the tendency for the Heinrich-type event to follow closely after a quite low minimum of insolation.

The history of glaciation as inferred from the $\delta^{18}O$ proxy for ice volume over the last three million years can also be understood in terms of the role of the Mediterranean in the neo-Milankovitch model. The enlargement of the dimensions of the Strait of Gibraltar that began 4.8 million years ago, when the isolation of the Mediterranean ended, probably continued over the last three million years since the Isthmus of Panama rose out of the sea. Three million years ago, with a smaller cross section of the Gibraltar strait, the exchange currents would have been much smaller than today. Although the outflow then was also smaller, it would have carried approximately the same amount of excess salt to balance the Mediterranean salt budget. Therefore, the smaller outflow would have been quite saline and much more dense than today's outflow. The greater the density difference between two liquids, the more difficult and slower is the mixing. Consequently, the denser outflow would have sunk downward into much deeper levels of the North Atlantic than today before mixing to buoyant equilibrium. The outflow would therefore have contributed very little to the high-latitude salinity, as suggested on the right hand side of the schematic diagram (Fig. 29) of excess salt reaching the Nordic Sea. As a result, the initial $\delta^{18}O$ fluctuations and ice-volume changes three million years ago were quite small.

As the Strait of Gibraltar enlarged, the salinity and density of the outflow decreased, buoyant equilibrium of the outflow was attained at shallower depths, and the delivery of Mediterranean salt to the high-latitude sea surfaces became more effective, with an eventual increase to the maximum point that is suggested in Figure 29. In glacial times during insolation minima at 25°N, this would have promoted the intermediate-level deepwater formation in the Labrador Sea and the moderately warmer high-latitude sea-surface conditions, leading to optimum conditions for ice-sheet accumulation.

It may be that, near the top of the curve in Figure 29, the Eurasian ice sheet reached a maximum that is dimly seen in the glacial terrain of eastern Europe as an extension of glaciation into the headwaters of the Volga River, an extension that has not since been attained. This earlier blockage of Siberian rivers to the Arctic and the resulting meltwater flood into the Mediterranean probably caused a larger anomalous mid-Pleistocene deglaciation of longer duration than that of 140,000 years ago. The effect on the Laurentide and Antarctic deglaciations would also have been more extensive, and could have caused the sealevel rise to the +20 m level reported by P. Hearty et al. in 1999. An orbital configuration at 557,000 yr BP with insolation almost identical to that of 150,000 yr BP (Fig. 55) implies a date sometime after 550,000 yr BP or slightly later for blockage of the Siberian rivers and start of the flooding of the Mediterranean with meltwater after the overflow via the Turgai Pass. This earlier and longer anomalous deglaciation apparently created the large coral-reef terrace known as the Second High Cliff on Barbados (Fig. 11), which is approximately of that age and is larger than the First High Cliff. In the later anomalous deglaciation, the Eurasian ice was somewhat less extensive, the sealevel rose only to +7.4 m instead of +20 m, and the First High Cliff was less extremely developed.

The orbital insolation differences (Fig. 55) were practically identical in the two cases, so why were the events different ? A higher anomalous sealevel of +20 m in the earlier event is consistent with a longer interval of conveyor-belt shutdown due to a thicker and more extensive amount of glacial ice in Eurasia. Yet, prior to 550,000 yr BP, the time available to build up the Eurasian ice after the previous interglacial high in the $\delta^{18}O$ was only one precessional cycle. On the other hand, prior to the later anomalous deglaciation the interval of ice buildup was about two precessional cycles from the previous interglacial about 200,000 yr BP. The rapid earlier buildup would be consistent with a more effective delivery of Mediterranean salt to the high latitudes, as would be expected if the Strait of Gibraltar had enlarged enough to optimize the delivery of salt suggested at the maximum of Figure 29. Since that time further enlargement of the strait could have reduced the delivery of salt to the high latitudes, thus causing a more-zonal atmospheric flow over the North Atlantic resulting in a slower buildup and a smaller thickness of Eurasian ice with twice as much time needed to block all the Siberian rivers.

In summary, the evidence and network of connecting arguments point to the Mediterranean Sea and its outflow at Gibraltar as the main factor in major climate fluctuations over the last three million years since the emergence of Panama. Today, excess salt from the Mediterranean outflow combines with salt from a branch of the Gulf Stream to enable strong deepwater formation in the Nordic Sea, yielding a relatively stable interglacial climate. Without the Nordic Sea deepwater formation, we have an unstable glacial climate. In glacial times the North Atlantic Drift becomes much weaker and less important, while the Mediterrannean outflow, driven by monsoons, becomes a controlling factor in North Atlantic oceanic and atmospheric circulation. The outflow variation is then quite often a key factor in the advance or retreat of northern ice sheets. A timely question is: Will the Mediterranean likewise play a role in the climate changes stemming from mankind's influence on the world environment ?

Before getting out our crystal ball and looking into the future in the last chapters, a few parting philosophical remarks may not be out of line. Those who feel that all relevant questions should be answered before accepting the neo-Milankovitch hypothesis may not be comfortable with this model of climate change. Many facts from the geological record that are not yet widely accepted are combined with reasonable inferences to form the complex picture of ice-age climate oscillations. This inductive approach to increase our insights into the causes of climate change is defensible because certainty is difficult to attain. Considering the complexity and the chaotic nature of Earth's climate system, it may be impossible to get reliable numerical model results or quantitative evidence from the earth itself to confirm beyond doubt all the elements of the neo-Milankovitch hypothesis, and our knowledge may ultimately be dependent on these logical inferences. But the controversial elements will surely be investigated, because in the earth sciences the last word is seldom said.

16

Threshold clues from the Norse sagas

Evidence from the geological records discussed in Chapter 12 points to a warming of the Labrador Sea that triggered the last ice age 120,500 years ago. The cause of the warming was traced all the way back to the decreasing northern subtropical summer insolation at 25°N latitude by way of its effect on monsoons and Mediterranean salinity and outflow. If orbital effects on the insolation are the key to the next ice age, when do the astronomers say it might begin ? Well, the astronomers wisely report only their calculations and leave the predictions to the earth scientists. However, the prediction based on classical Milankovitch thinking is not the same as one made using the neo-Milankovitch concept involving the Mediterranean Sea and low-latitude summer insolation. This is because Earth is now entering another 20,000 year interval in which the precession effect in the north opposes the polar-axis tilt. Consequently, Northern Hemisphere summer insolation is falling in the higher polar latitudes, while low-latitude insolation is rising. Figure 56 shows these contrary trends.

John and Kathryn Imbrie in the book *Ice Ages: a Mystery Solved* proposed an answer to the question in the classical Milankovitch context in which the emphasis is on the high-latitude insolation. After all the fossil fuels have been used up and atmospheric carbon dioxide returns to a more normal level 2000 to 3000 years from now, the high-latitude cooling would become apparent at somewhat

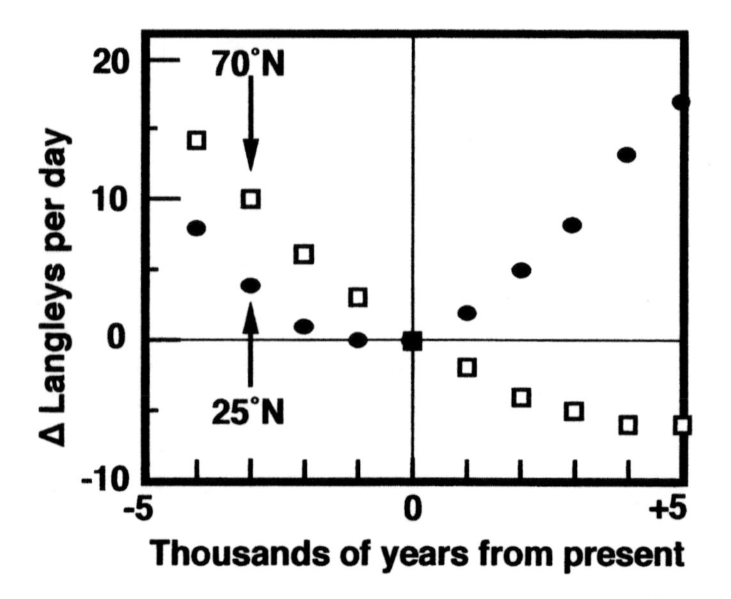

Figure 56: Insolation trends at high and low latitudes. Caloric summer insolation differences from the present day values (Fig. 6) through present. Tabulated values from Berger (1978).

lower insolation than present. The aridity associated with colder climate would slowly set in as insolation continues to fall, and 20,000 years from now when precession and inclination effects once again come together, the earth would be well into the next ice age. They state one caveat - the assumption that the increase of carbon dioxide in the atmosphere occurring in the the fossil fuel era will not cause a fundamental change in the climate system.

However, Walker and Kasting in 1992 predicted CO_2 levels at about 2000 parts per million in the atmosphere 400 years from now when most of the coal, oil, and gas have been consumed. This concentration is more than five times that of today, and is about 30 times the increase in CO_2 that has occurred since the industrial age increase began. This implies a very large climate warming. Therefore, the Greenland and Antarctic ice caps may melt, and with sealevel about 65 meters higher than today and all the coastal areas of the world flooded, a return to "normal" Pleistocene conditions would not be likely.

The neo-Milankovitch model of climate change predicts a much more rapid glaciation, but this is because humanity's alteration of the Mediterranean environment plays a critical role, as discussed in the next chapter. The neo-Milankovitch model also has quite interesting historical implications. Figure 57 compares the insolation trend of summer insolation at 25°N during the initiation of the last ice age with insolation through the modern era. It is plain to see that about 1500 years ago low-latitude falling insolation flattened out near the level that triggered new glacial growth 120,500 years ago, and has now just started to rise. There is a ±500 year uncertainty in the timing of the trigger, so the trigger level of insolation could be slightly higher or lower than shown. There are also "noise" fluctuations in the climate system that can shift the trigger level up or down. We see such fluctuations in the modern North Atlantic oscillation discussed by J.W. Hurrell in 1995. In Scandinavia winter air temperatures can average 3°C warmer (or colder) while at the same time Labrador temperatures average 3°C colder (or warmer) than normal. Regardless of just where the trigger is, the new ice age has not yet begun, but the insolation level indicates that we might be near the triggering threshold, with the possibility that an unforeseen climate fluctuation could trigger a new major glaciation.

Is it possible that in our historical records there is some indication that the climate is near the threshold ? The severe winters in Europe and eastern North America during the Little Ice Age 150-300 years ago are sometimes cited as the type of climate change that would occur at the threshold of the next ice age, and there are perennial snowfields in the higher elevations of Baffin Island that were more extensive then and that seldom melt today. If Milankovitch cooling were initiating the transition, the Little Ice Age cooling could be an excellent indication. But as we have seen, the geological records tell a different story in which the first signals of change are actually warmer sea-surface temperatures associated with higher salinity in the high latitudes of the North Atlantic.

In the latter part of the first millennium, an interval of warm climate began in the northern North Atlantic that lasted until about 1270 AD. It facilitated a westward expansion of the Norse pastoral peoples from the over-populated Scandinavia of that time, an activity recorded in the Norse sagas. About 870 AD Norse settlement in Iceland began and the land there was soon occupied. The climate was favorable for pasturage, and small birch trees were plentiful when the first settlers arrived. The Norse were consummate explorers and they soon became aware of the much larger island of Greenland a short sail to the west.

Eric the Red, a rather quarrelsome Icelander, was exiled from Iceland for three years by his peers because he killed a man. He spent his exile time exploring the southern and southwestern coasts of Greenland and found new land for

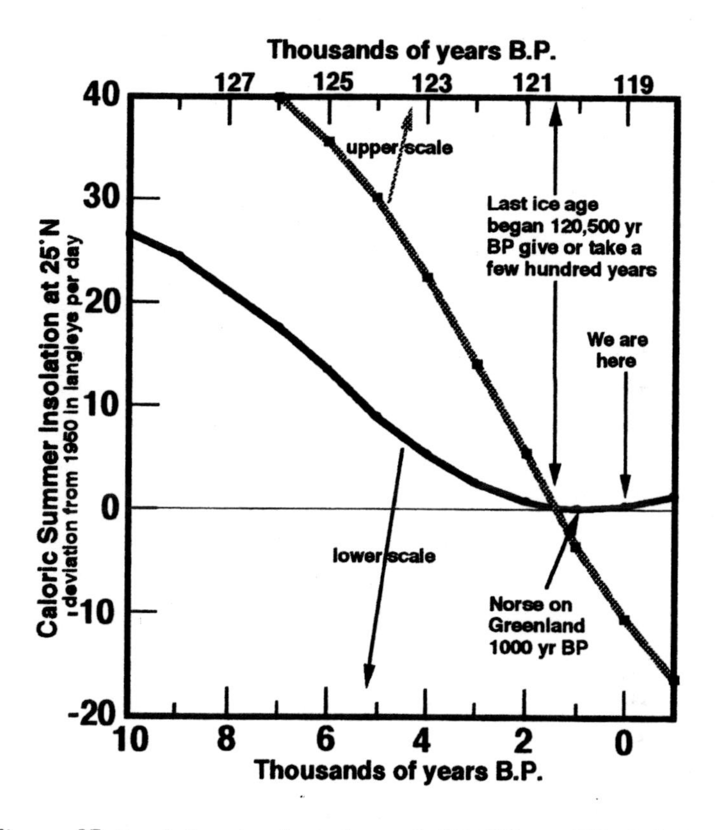

Figure 57: Insolation trends today and 120,000 yr BP. Caloric summer insolation at 25°N latitude when the last ice age began compared with the trend through the Holocene. Today's insolation is close to the level of 120,500 years ago, and suggests that we are at the threshold for new ice-sheet growth. However, our climate has not yet crossed the threshold. Tabulated values supplied by Berger (1978).

settlement in some of the fjords west of the southern tip, now known as Cape Farewell. Upon his return to Iceland he advertised his discovery, and called the new land "Greenland" to attract settlers from the now fully populated Iceland. He was apparently a good salesman, because he led a flotilla of Icelanders west to Greenland about 985 AD and settled in the area of Julianehab, a short distance west of Cape Farewell. The climate was favorable for pastoral farming and the Greenlanders prospered. By one estimate, the fjords on the south coast may have been 5°C warmer then, based on the saga story of a Norse chieftain who swam across the fjord to get a sheep to bring back for a feast in honor of a guest. This implies a warmer or stronger Irminger Current and West Greenland Current than today.

If the fjords were in fact 5°C warmer, this would imply a greater freedom from sea ice throughout Baffin Bay. This raises an intriguing question, could the Norse have sailed around Greenland ? In Helge Ingstad's 1969 book: *Westward to Vinland* there is an old map, dated 1599 (Fig. 58), that was found in an archival collection in Hungary after World War II. The map was drawn by a Jesuit cartographer and shows Greenland as an island with a realistic outline. More recently (Ingstad and Ingstad, 2001), it was suggested that the map is a later forgery, possibly of the 18th century, based to a large degree on linguistic anomalies. But modern linguistic standards applied to ancient texts are notoriously unrealistic because of the lack of standards of linguistic practices in early times, and it is recognized that early cartographers utilized information from various sources and cultures. Therefore, it is likely that the Jesuit cartographer in 1599 used his most current information as well as material he found in an older source map, now lost, that was drawn by an earlier church cleric, as Ingstad suggested in 1969.

The presentation of Greenland as an island would have been anomalous in 1599. All the maps drawn by explorers

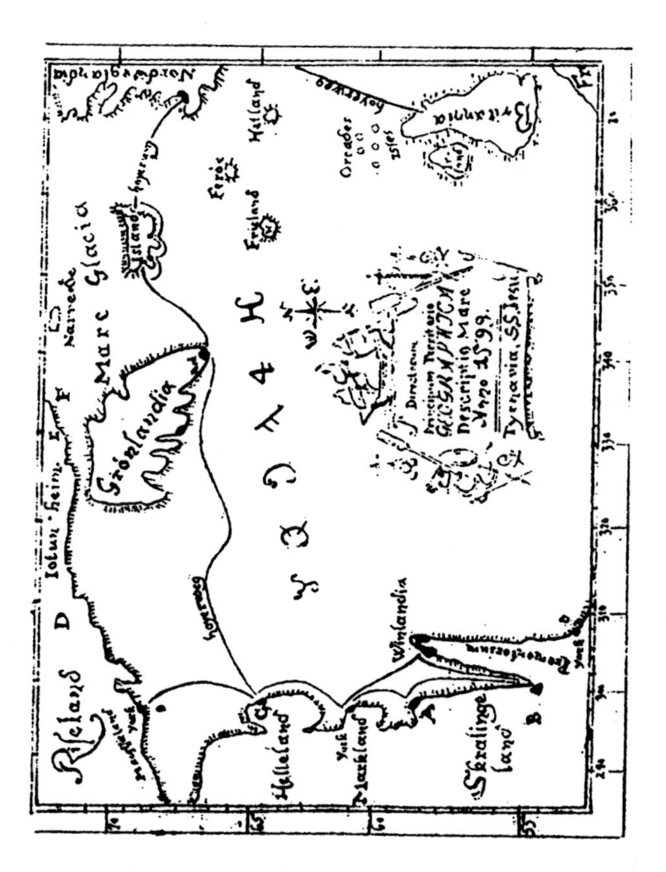

Figure 58: Norse trading routes on an old Jesuit map of 1599 AD from Helge Ingstad's book: *Westward to Vinland.* The Jesuit cartographer used information from a much older map (now lost) to show trading routes used by the Norse, probably in the 12th century. None of the other maps drawn in the 16th and 17th centuries show Greenland as an island because heavy sea ice prevented the post-Columbus explorers from sailing to the northern regions. In this figure, most of the blemishes that accumulated in the background as a result of the age of the original document have been removed.

of those times show Greenland attached to the continent because they were unable to penetrate the heavy sea ice that blocked the channels between Greenland, the Queen Elizabeth Islands, and the continent to the southwest. It was not until 1891 that Greenland was shown to be an island by an expedition led by (Admiral) Robert Peary.

The Jesuit order, founded in 1534, had a strong missionary interest in exploration. However, the Jesuit map is not a general map of Norse sailing routes of early times because some well-traveled routes are omitted. For example, routes are not shown between Scotland and Bergen in Norway, between Scotland and Iceland, or between the eastern and western settlements on Greenland. Instead, the map is apparently a faithful map record of a two-year voyage that began in Europe and, for the most part, followed known trading routes. The original cartographer was most likely Bishop Eirik Gnupsson who, according to the Icelandic Annals, visited Vinland about the year 1121 (Skelton et al., 1965, p. 223), and it is said that he wintered over there. The bishop probably traveled with a trader on one of the merchant ships that were in use in the 12th century. The source map for Figure 58 may have been the only surviving documentation of his voyage, although he must have also composed a written account for his superiors. It is of great interest to reconstruct and retrace the bishop's voyage.

After traveling to Scotland, he sailed to Oslo, and went overland to the port at Bergen on the coast of southwestern Norway. By that time it was probably midsummer. There he joined a ship's company and sailed from Bergen to Iceland, where they traded in a leisurely fashion at four ports. They then sailed onward to the eastern settlement on Greenland, designated as a york, probably at the Norse headquarters at Brattahlid at the head of a long fjord 80 km (50 miles) from the open sea. A "york" is the Anglicized version of the Norse word "jorvik," a trading settlement located on a stream inlet somewhat in from the sea. After trading at Brattahlid they set out along the coast, sailing with the West Greenland Current. Their course soon turned westward and then northwestward to compensate for the southward flow of the Canadian Current, enabling them to reach the port in Helleland. This site was likely in the sheltered Eclipse Channel on the south side of North Aulatsivik Island. It was late in the sailing season, and at their next port, the Markland york, they spent the winter.

There are intriguing clues in the Jesuit map of Markland that can be matched with features on modern Canadian topographic maps. The Markland york is located at the south end of a large bight (Fig. 58), which is filled with islands (Fig. 59). At the northwest corner of the bight on the Jesuit map is a small tear-drop-shaped bay. This is Tikkoatokak Bay, which connects to the bight through a narrow inlet on the Canadian map. Most of the inlets in the bight have shallows that would not provide good mooring sites due to the large amplitude of the tides along the Labrador coast. The only inlet with deep mooring sites that meets the definition of a york is Flowers Bay, which has all the advantages a Norseman could desire. It is a snug inlet about fourteen miles long, with a substantial river entering its west end. Deep mooring sites are found near the river mouth and adjacent to elevated ground that would have been ideal for houses or shelters. Nearby brooks would have supplied fresh water. Salmon fishing, goose hunting, and exploring the island-filled bight probably

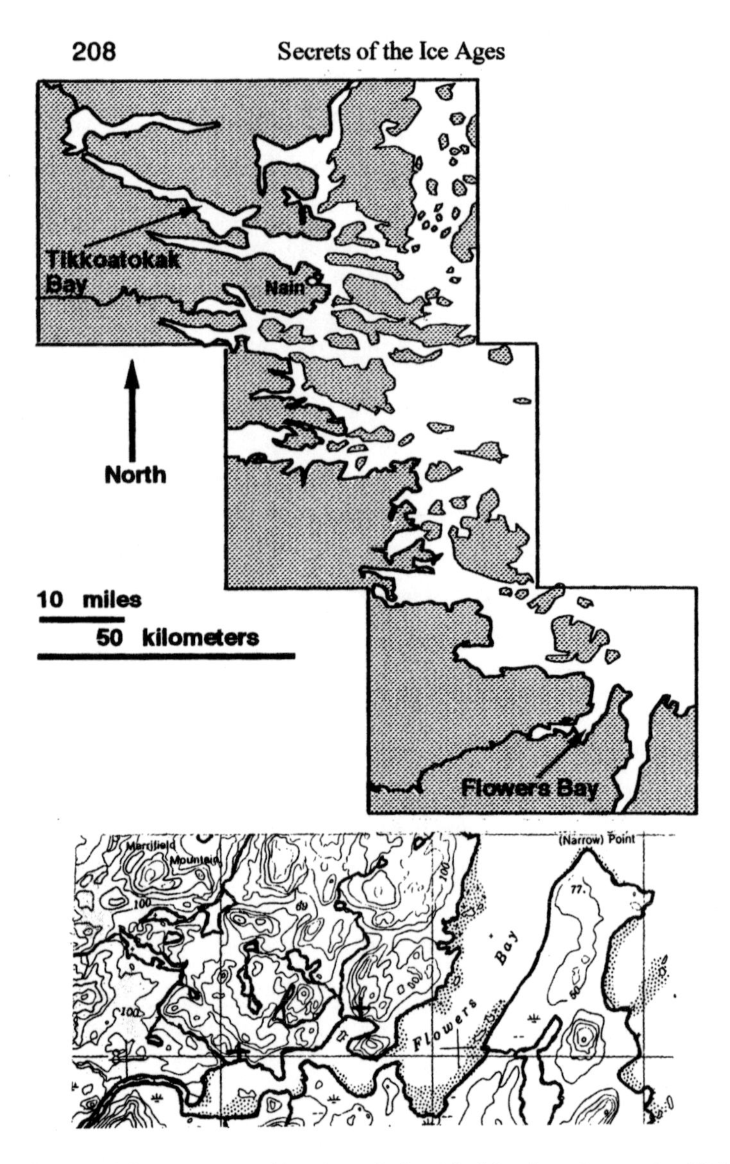

Figure 59: The probable site of the Markland york on the Nain bight. The tear-drop bay (Tikkoatokak Bay) and the york where the bishop and the Norse traders may have wintered over during a two-year trading voyage to Vinland in the 12th century are identified on Canadian topographic maps. Crosses on the detailed map of Flowers Bay indicate likely locations of shelters.

occupied their fall and winter days, and the storms that would have made life less pleasant at the exposed L'anse aux Meadows Vinland site on the tip of the Newfoundland promontory would not have bothered the bishop and his friends who were wintering on Flowers Bay.

The following spring they sailed southward along the coast to the site marked A (Fig. 58), probably on the modern Sandwich Bay. From there they crossed the Strait of Belle Isle and traveled down the Newfoundland coast to site B, which was likely on the modern Port au Port Bay. Spring weather in Newfoundland often consists of foggy and rainy days, and lack of visibility might explain the cartographer's mistaken assumption that the coast line from Sandwich Bay extended directly south to the Port au Port Bay, which is enclosed by a long peninsula on the west side. With trading completed there, they sailed the coast northward to stop briefly at one of the settlements on the promontory, then again at the Markland york at Flowers Bay. Their next stop was the york on Baffin Island, which was labeled Rifeland on his map. It is also designated as a "Mocgtjaland," that is: a "Mosquitoland" when translated phonetically with a soft *c* and soft *j*. They probably arrived at this york early in the summer when the mosquitoes and black flies are at their worst, and would certainly have impressed a visiting European man of the cloth who had never before seen an Arctic summer.

Beyond that mosquito-plagued site, however, their route becomes a real puzzle. One would expect them to have taken a trading route directly back to southern Greenland, but they did not because this route is not shown on the map. And they certainly did not return to Helleland and then retrace the hoyerweg against both the Canadian

Current and the West Greenland Current. But there would have been no trading route northward around Greenland, and if they sailed that way, the traveling bishop would not have shown it as such a route. Distances on the map would have been strongly influenced by the time required to sail between ports. The true distance from the Mocgtjaland york to Iceland is twice as great to the north around Greenland as in the more direct route to the south. Yet in Figure 58 the distances presumably perceived by the bishop are about equal. This is what one would expect if they sailed around northern Greenland in not many more days than they needed to sail from Iceland to Helleland against the westerly winds. If they did sail around Greenland, they were not the first to do so, because the Grönlandia outline suggests that the coasts were well known to the Norse, which implies that all the Greenland coasts were bordered by seas that were warmer and much more ice-free than today.

There is an intriguing bit of evidence on the map that hints that our bishop may indeed have circumnavigated Greenland. At the northern end of Baffin Bay on the old map (Fig. 58) the area is designated as "Iotun heim." On Stefansson's Skalholt map of 1570 AD (Skelton, 1965), Greenland is shown attached to the continent, consistent with all of the other maps drawn at that time, based on the post-Columbus explorations that could not penetrate the perennial northern sea ice. On Stefansson's map, however, the corresponding northern area is named "Iotun heiman", which apparently is an old plural form, ending with *an*. With an interchanged *I* and *J*, this resembles the modern Jotunheimen, a name that translates as "Home of the Giants" and is attached to well known mountains in western Norway. There is a mountainous area on Ellesmere Island that could have originally received this name, but why would the term "giant" be attached to such mountains, and why would the Jesuit map, based on much earlier history, have used the singular form ? "Iotun heim" is "Home of the Iotun" where the "Iotun" is a population of individuals, just as "Skralinge Land" is "Land of the Skralinge," or in modern terms, "Land of the native American." But who were the Iotuns ?

Recall that our traveling bishop was an educated man, and would have known something of Greek mythology. In one Greek myth, Io was a beautiful maiden in a love triangle with the god Zeus and his wife Hera. To protect Io from Hera's jealous machinations, Zeus transformed her into a heifer (a young cow). As a heifer she wandered all over the world, eventually returning to Egypt where she was transformed back to her womanly form. The long-haired Arctic muskoxen that live along the coasts of Baffin and Ellesmere Island could have been observed by the bishop. The muskoxen greatly impressed the early explorers, and were known as "giant cattle" that were very "difficult to take," according to a Latin note on Stefansson's map. The name "Home of the Giants" may originally have been merely a shortened version of "Home of the giant cattle." So somewhere in the vicinity of the mountainous area on Ellesmere, the Bishop probably had first-hand contact with the muskoxen, and he named the area "Home of the Iotun" after the mythical heifer, Io.

The Norse sagas leave us with one more climate indicator: heavier precipitation. H.R. Holand's translation of one of the sagas tells the story in his book: *Westward from Vinland.* Shortly after the first wave of Norse settlers arrived in the fjords of southern Greenland, a young Norse trader by name of Bjarne Herjulfson returned to Iceland from a trading voyage to Norway, intending to spend the winter with his father in Iceland. When he arrived in Iceland, he was surprised to learn that his father had emigrated to Greenland. Although he had never been to Greenland, he was determined to winter with his father, so he obtained sailing directions to his father's new home, collected his crew, and departed (Fig. 60).

Among all the Norse sailors, there was no one more expert or daring than Bjarne Herjulfson. Without even a primitive magnetic compass for navigation, with only verbal directions to guide him to a Greenland destination that he had never seen, and with the end of the summer sailing season close at hand, he proposed to sail to his father's new home, if his crew would accompany him. That they consented to the voyage is testimony to Bjarne's reputation as an expert seaman.

Soon after setting sail from Iceland, they encountered a severe storm and were driven off-course far to the south, the saga says. We can infer Bjarne's thinking and his decision in setting a new course when the storm passed and the clouds cleared away. Open-ocean navigation of that time depended on sailing east or west on known latitude lines, which were determined by measuring the altitude of the north star or the noontime sun and applying a correction to the sun's elevation that depended on the time during the summer. Although Bjarne's sailing directions included the latitude of his father's new home, he was unfamiliar with the Greenland coast. Therefore, he could not sail northward to his father's location with any confidence because he did not know if he was east or west

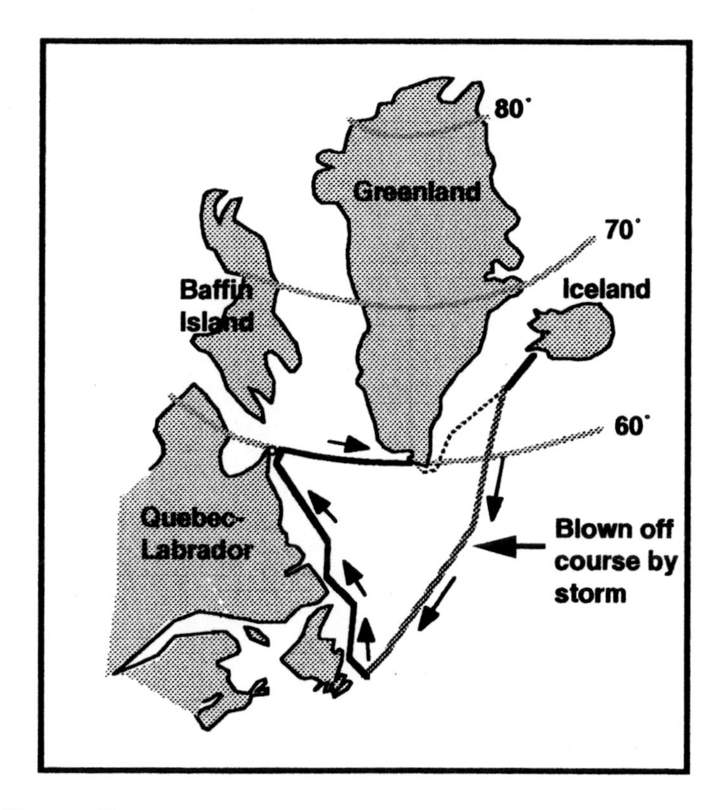

Figure 60: Bjarne Herjulfson's voyage of discovery of America. See text. Based on a Norse saga translated by H.R. Holand (1940). Dotted line - his intended course. Shaded heavy line - course during the storm. Heavy black line - his course after the storm, chosen to be west of his father's home before sailing east on the desired latitude line.

of that destination. However, with a sight on the sun, he now knew how far south he had come. His decision was to sail northwest in what he thought was an empty ocean until he reached the desired latitude, where he would certainly be some unknown distance to the west of Greenland. He would then sail due east with the winds at his back to reach his father's location. This he did, and in doing so, he became the first Norseman to sight North America.

On the first sailing day after the storm, the saga says, he sighted land now believed to have been Newfoundland. It did not resemble the description of Greenland, so he pulled far away from the land and again sailed northwest. A few days later they sighted land again, a wooded coast probably on Labrador, but still not Greenland and not the right latitude. His men wanted to go ashore, but Bjarne refused. His only objective was to get to Greenland. So they pulled away and again headed northwest. A few more days later, he sighted a barren and mountainous coast and noted a high mountain "with ice upon it." As they sailed by the mountain, they "saw that it was an island." Then he turned the bows away from the land and sailed with a wind so strong at his back that they dared not use all their sail. On the evening of the fourth day after leaving the island mountain, they arrived directly at his father's house.

The method of navigation by latitude line dictates that the mountain with ice on it could only have been 3100 ft-high North Aulatsivik Island, and, looking at detailed Canadian topographic maps, it is clear that if you come from the south and sail on passed the mountain and look back, you can see that it is certainly an island. Bjarne's mountain is located near the northern tip of the Labrador Peninsula, and on the same latitude as Cape Farewell and the tip of southern Greenland. Today, snow does not remain through the summer season on the coastal mountains of Labrador. But a thousand years ago at the edge of a warmer Labrador Sea, new glacial ice or a perennial snowfield was accumulating on the mountain there, as would be expected at an ice-age threshold.

17

Handwriting on the wall

Despite the ice or snow reported by Bjarne Herjulfson on the Labrador island mountain on that historic summer over a thousand years ago, no ice age has yet begun, and, living in our new CO_2 greenhouse, it would not seem likely. Nevertheless, in 1992 Miller and de Vernal in a letter to *Nature* posed the question: "Will greenhouse warming lead to Northern Hemisphere ice-sheet growth ?" The thought behind the question is quite reasonable. The Arctic regions of northern Canada are quite cold, even in summer. Warming of low latitude regions will increase evaporation rates, and the added moisture circulating in the atmosphere could increase snowfall in Arctic Canada and initiate large-scale glaciation, even in a generally warmer climate. Once established, strong feedback cooling and precipitation patterns caused by the presence of the ice sheet itself would make the glaciation difficult to remove. We now know more about what is required to initiate Canadian glaciation, and with the arguments of the earlier chapters in mind, the question can be asked again in more focused terms: What are the specific changes in the climate system due to man's activities that are likely to initiate new Canadian glaciation, and when will that event occur ?

A valuable perspective on this question can be gained by looking again at Figure 57 and comparing the critical low latitude insolation of 120,500 yr BP, when the last ice age began in Canada, with today's insolation trend. Here we see the steadily falling insolation 123,000 years ago during the warmest part of the Eemian interval. It continued to fall, passing below the ice-age initiation threshold to a minimum at 115,000 yr BP. The insolation minimum guaranteed a strong Mediterranean outflow and a large supply of excess salt to the higher latitudes and the Labrador Sea, thus increasing the probability of deepwater formation there or in Baffin Bay. The warmer replacement water reaching Labrador brought storm tracks to the north, and caused rapid Laurentide Ice Sheet growth. The trend of insolation in the recent past and near future shows that the insolation level fell to the threshold level at which glaciation began about 120,500 yr BP, but the curve flattened out and now has just started to rise. No new ice age has been initiated, but we can see the threshold climate signs in the modern African monsoons. Today, monsoons only extend northward into the Nile headwaters, but 9000 years ago, when the Sahara was an inhabited grassland, monsoons extended north as far as central Egypt (Fig. 51). Frequent droughts in the sub-Sahara testify to the limited rainfall of modern times, and the Near East is more arid now than depicted in early historic times.

The Norse sagas and geological records suggest that climate conditions a thousand years ago were perilously close to the threshold for new ice-sheet growth. However, the threshold was not crossed, and we, or rather our ancestors, may have avoided a new ice age by the narrowest of margins. Some unexpected fluctuation apparently aborted the oncoming climate switch. Perhaps it was a decrease in the intrinsic output from the sun. The Sporer sunspot minimum, a time of weaker solar activity discussed by M. Stuiver in 1980, may have ended the Norse prosperity around 1300 AD when the climate turned colder. Or maybe it was caused by the "bi-polar seesaw" (Appendix B). Regardless of the details, the trend of falling low-latitude insolation as determined by orbital change has now been reversed. If all the other variable climate factors had remained unchanged, the now slowly-rising insolation should begin to increase the monsoon strength. The humidity over the Mediterranean and Near East should rise, the Nile flow should increase, the outflow at Gibraltar should diminish, and it could be that Earth has been spared another 100,000 year-long ice age. But wait ! Like the Old Maid in the childs' card game, the threat has returned, and when the game finally ends we may yet be holding the Old Maid card.

What has gone wrong ? Well, European oceanographers report that the salinity of the Mediterranean Sea is increasing, according to an article in Science (Fig. 61) in 1998. An increasing Mediterranean salinity means an increasing outflow at Gibraltar, and a warmer Labrador Sea. And we know that a warmer Labrador Sea triggers ice ages. Two factors are responsible for the Mediterranean salinity increase. One factor is the sea-surface temperature increase due to global warming. In the horse latitudes, where stratospheric air sinks and dry climates and deserts are the rule, the global greenhouse warming will certainly increase evaporation losses, but a compensating increase in precipitation is quite unlikely. The Mediterranean Sea should therefore experience greater net evaporation losses and these can be estimated from the increase in vapor pressure as the sea-surface temperature rises. A second factor is the widespread use of irrigation water taken from all the rivers that empty into the Mediterranean, as discussed by Rohling and Bryden in 1992. Most of this irrigation began in the modern era of dam construction within the last 50-100 years. Even the flow of rivers into the Black Sea has been greatly reduced to provide irrigation water. The estimated equivalent loss of freshwater flowing from the Black Sea to the Mediterranean is about 700 m^3 sec^{-1}. This loss is, however, small compared to the decrease in Nile River discharge to the Mediterranean.

Since the dawn of history, Egypt has depended on the Nile for its national existence, for Egypt sits in a desert of the horse latitudes and its only source of water is the Nile, fed by the African monsoons. In recent historic times, the water usage for Egyptian irrigation has greatly increased. Prior to 1968, about 35% of the normal average flow of 2800 m^3 sec^{-1} was withdrawn for that purpose and 65% reached the Mediterranean. In 1968 the Aswan High Dam was completed, the fraction of Nile water reaching the Mediterranean fell to about 10%, and the loss of flow increased to about 2500 m^3 sec^{-1}. The remaining flow may be further reduced as the Abu Simbel diversion project to pump water out of Lake Nasser is completed, according to a 1994 paper by Gasser and El-Gamal. The loss of Nile freshwater represents about a 20% reduction in the total river input to the Mediterranean Sea, as estimated relative to conditions before much irrigation usage occurred. It also represents almost 9% of the net hydrologic deficit of the Mediterranean Sea.

If it requires about 120 years to achieve 63% of a new salinity equilibrium, as discussed in Chapter 10, then clearly due to irrigation losses alone the salinity should be increasing, and, at the 2000 m-depth in the western Mediterranean over the last 40 years, it is (Fig. 61). Will

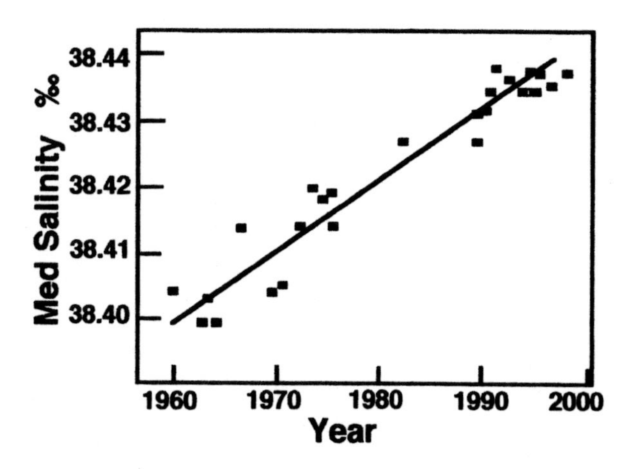

Figure 61: Rising salinity in the deep Western Mediterranean since 1960. Modified from data published as a news item in *Science,* vol. 279, p. 483-484, 1998. Data reported in a European conference by Béthoux and Gentili.

this increase cause a corresponding increase in the outflow at Gibraltar ? The answer is yes. Although the Mediterranean deepwater temperature is increasing, which reduces its density, greenhouse warming is likewise increasing the temperature of the inflowing Atlantic water, probably even more so. Therefore, the temperature effects on the difference between densities of the upper inflowing water and the outflow at Gibraltar should cancel out, and the salinity effect will stand alone as an increase in the density difference at Gibraltar.

It is of great interest to predict the increase in the outflow at Gibraltar caused by irrigation losses and evaporation losses due to greenhouse warming. The main factors forcing the outflow are the density differences between outflow and inflow, and the westward slope of the interface between inflow and outflow over the sill in the western part of the strait. This driving force must overcome the effect of the drag on the bottom and the drag at the interface between flows. In addition it also must overcome the pressure gradient that drives the inflow above. The inflow, however, has a cross section about twice as large as the outflow. The picture becomes even more complicated when one considers that, if the density difference increases, the interface slope will also increase. The slope increase tends to compensate for the negative effect of the increase in the inflow pressure gradient that must occur to keep the inflow and outflow almost equal, with a stable sealevel in the Mediterranean. These are complex factors, but it appears to be a reasonable approximation to assume that the outflow is simply proportional to the salinity (density) difference between the two flows at the sill. Today this is about 1.7‰, which is slightly less than the 2.2‰ open ocean difference because of countercurrent exchange of salt across the interface between the two flows. The percent increases in future years can then be estimated using a stepped computer integration by calculating the annual salt added to the Mediterranean, assuming it is perfectly mixed, and then calculating a new salinity and outflow for the next year.

Figure 62 shows the result of this "back of the envelope" computer calculation. It predicts the Mediterranean outflow change due to both the loss of 90% of the Nile River water by irrigation and the added effect of greater evaporation losses as greenhouse warming increases the temperature and the vapor pressure at the sea surface. The 90% loss is assumed to have started in 1968 when the

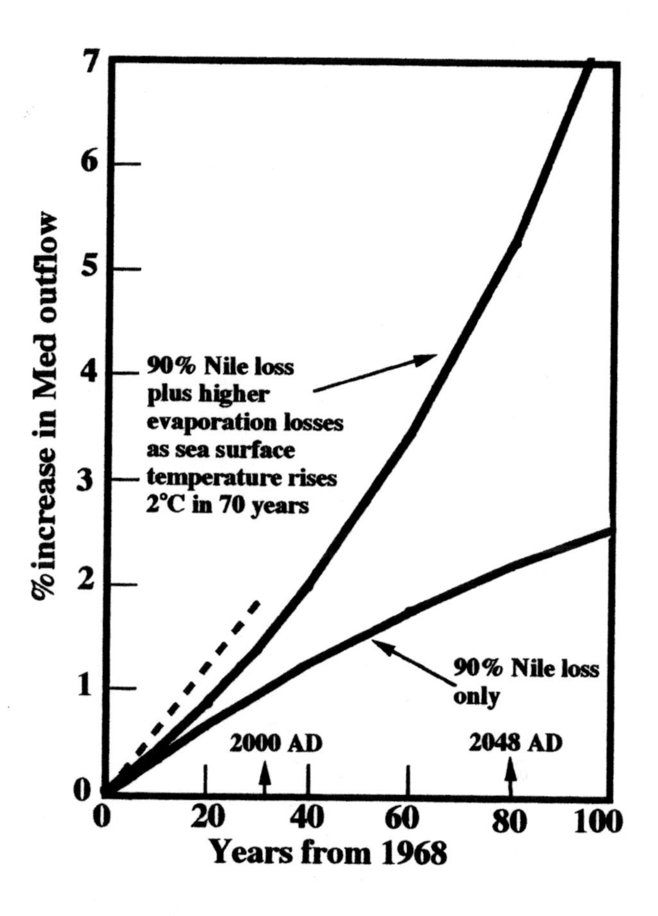

Figure 62: Estimate of Mediterranean outflow increase using a simple model and integrating by annual steps. The predicted increase in the outflow due to Mediterranean Sea surface warming over the next seventy years exceeds the effect due to the loss of Nile River discharge caused by irrigation. Other significant irrigation losses have been neglected in this calculation. A similarly increasing outflow is thought to have initiated the last ice age about 120,500 years ago. The dashed line is the increase in outflow that would be expected from the observed salinity and density increase of Figure 61 if the temperature increases in the exchange currents are equal.

Aswan High Dam was completed. This is not quite correct, of course, because modern irrigation losses began somewhat earlier. But the 700 m^3 sec^{-1} reduction of river water entering via the Black Sea has not been included in this calculation. A better estimate of the outflow increase should therefore come in slightly above that shown in the figure. The dashed line based on the salinity (density) increase suggests that the estimate is a reasonable one.

It is important to note that the amount of excess salt carried out of the Mediterranean into the Atlantic increases at a rate twice as fast as the outflow itself because both the salinity and the outflow are increasing. Consequently, within the next 70 years, if the outflow at Gibraltar increases by 5%, the excess salt entering the North Atlantic would increase by as much as 10%. An outflow of excess salt that increases by 10% at Gibraltar would represent an increase of about 4% in the excess salt reaching high latitudes if salt from the Gulf Stream is included in the total. However, the fractional high-latitude salt increase would probably be much greater than 4% due to the nonlinearity suggested by Figure 29. In addition, if much of the excess salt increase goes directly into the Norwegian Sea and on into the Arctic ocean, it would be quite effective in reducing the extent and thickness of the packice there. Although the predicted outflow increase of Figure 62 is quite approximate, it is clear that the trend is not negligible if we are near a threshold for new glaciation. Man's activities are rapidly increasing the outflow at Gibraltar, just as falling insolation did much more slowly 120,500 years ago. This may be the handwriting on the wall, and there is more.

Although the Greenland fjords are not yet as warm as

they were a thousand years ago, it is worth noting that in the summer of 1998, news reports stated that for the first time in the records since the Titanic sank in 1912, no icebergs appeared in the shipping lanes to Europe, and the Labrador Sea was 2°C warmer than usual. Two years later, Krabill and colleagues reported that the higher-altitude ice of the Greenland Ice Sheet in the southwest sector had thickened at an rate of about 27 cm per year over the last two decades, while the ice in the southeast sector had thinned. Is this a fluctuation, or is the shift of storm tracks from the east side to the west side of Greenland beginning to occur, consistent with a warmer Labrador Sea ? In another recent report, Rothrock, Yu, and Maykut state that sonar measurements of the packice thickness in the Arctic Ocean made during submarine voyages show that the ice has decreased in thickness by about 40% over the last 30 years. In another 30 years it will have largely disappeared if the trend continues.

Is this trend caused by global warming or is it caused by a higher Arctic Ocean salinity due to salt that is carried into the Arctic by the Norwegian Current, and is reducing the vertical density gradient and increasing convective warming in the packice area ? Is the extra salt coming from the Gibraltar outflow ? Convincing answers to these questions based on measurements are not available and may never be. The noise fluctuations and chaos in the climate system may prevent us from making the desired accurate measurements. But these indicators have the attention of the earth-science community. If the working hypothesis is correct that the loss of Arctic Ocean packice was the final step in the storm-track switch to the western side of

Greenland 120,500 years ago, then the next few decades will see the northern North Atlantic climate pass rapidly through the interval of threshold warmth and into the mode of large-scale ice-sheet growth in northeastern Canada - unless, of course, the greenhouse warming effect stops the very ice age that it is helping to start !

We do live in a remarkable world where sophisticated technology enables us to measure the thickness change in the Arctic Ocean packice as the years go by, and a thickness change in the Greenland Ice Sheet of only a few tens of centimeters per year. We even measure winter sea-ice coverage on almost a daily basis. As of this writing, regular measurements of sea-ice coverage in the Arctic are made by the GOES polar orbiting satellite operated by NASA. Any one with access to the internet communication system can look at and print out the false-color displays of the day-by-day ice coverage in Baffin Bay and the Arctic Ocean at the internet site:

<www.dcrs.dtu.dk/ftp/ssmi/icemaps/latest/ice.html>

maintained by the government of Denmark. If the trends discussed here continue, everyone with internet access will be able to watch the Arctic Ocean packice area shrink away in the next decade or two, and watch the winter sea ice vanish in Baffin Bay as the next ice age begins.

http://hobbes.emi.dtu.dk/sea-ice/iwicos/latest/ssmi.comb.gif

18

An ice-age prologue

No person living today will ever really see an ice age. The wheels of nature turn too slowly. But if the symptoms of ice-age onset appear as predicted, many of us may see a prologue to the next act in this ancient drama of the Pleistocene. The evidence favors an abrupt switch from the world as we know it to a somewhat different world, where the climate may teeter on the edge, while it decides whether or not to slide inexorably into a new ice age, or merely into some super-warm version of the Cretaceous period, a hundred million years ago. To say that we don't know which will happen is correct. But with a little evidence and some logic we may be able to look ahead a half century with an educated guess.

The problem is that worldwide greenhouse warming caused by rising CO_2 concentrations in the atmosphere could conceivably prevent the onset of early ice-age conditions, or might reverse such conditions even after they become established. A relevant question is: Could modern General Circulation Models of the atmosphere account for the CO_2, and provide answers to the question of new and lasting glacial growth in Canada ? Possibly, but in view of the chaotic nature of the climate system, our confidence in the modeling results would not be great. Nevertheless, there is a more general result of such models that seems to be reliable. The models predict that the warming in winter in high latitudes and the Arctic will be 2-3 times greater than in low latitudes, possibly because in cold climates the air is dry and CO_2 makes up a greater proportion of the greenhouse effect, relative to the contribution from water vapor in the atmosphere. This polar warming will tend to reduce deepwater formation and slow the northward conveyor-belt transport of oceanic heat in the eastern North Atlantic.

The effect of carbon dioxide warming on the conveyor-belt oceanic flow was examined in the 1993 model of Manabe and Stouffer under the assumption that the atmospheric concentration will increase at the present compound rate of 1% per year over the next 140 years to a level four times present. At that future time, the predicted conveyor-belt circulation in the model decreases to only 30% of today's circulation. The model decrease is caused by winter warming in the Nordic Sea, which reduces the amount of deepwater formed there. The implications of this result are warmer winters and colder summers in Europe in the future. Short-term noise fluctuations in the circulation model have a peak-to-peak amplitude of over 10%, and the trend of diminishing conveyor-belt flow does not become evident in the first 25 years. However, air temperatures in the model show an immediate steady rise, and the rise as of this writing may already be a factor in the thinning of the Arctic Ocean packice and may contribute to its disappearance in the near future. Also, the increasing outflow of salty water at the Strait of Gibraltar was not included in their model. The shrinking Arctic packice and the 3% increase in Mediterranean outflow by 2025 AD suggested in Figure 62 may therefore be quite adequate to switch the climate system into the Canadian glacial-growth mode. The expectation is that the switch would be rapid because the onset of deepwater formation in Baffin Bay would enhance the temperature contrasts with cold land masses to the west that favor strong low-pressure systems with wind stresses that drive the saline West Greenland Current into the bay - another positive feedback example.

The additional deepwater formation in Baffin Bay would increase the excess salt carried into the high latitudes by the replacement water and would greatly delay the predicted decrease in the conveyor-belt strength. Meanwhile, the cold surfaces of the growing new Canadian ice sheets under often-cloudy skies adjacent to the warmer Labrador Sea would resemble conditions at 120,500 yr BP. This suggests that the much heavier precipitation on the ice sheet could triumph over a small increase in global greenhouse temperature for many decades, and would certainly delay the predicted loss of conveyor-belt oceanic flow if the new Canadian glacial ice sheets keep high latitude winters cold. Consequently, with the simplest back-of-the-envelope argument it appears a good bet that we may see the switch to ice-sheet growth before the conveyor-belt shutdown is able to occur. If the glacial growth begins, say, within the next 25 years, what are the effects that we might expect in the first few decades ?

On Baffin Island and western Greenland, people living on the coasts might be able to remain for a time under the favorable conditions of a warmer sea, but they would seldom see the sun. However, mining or any other economic activity in the interior of Baffin Island would probably stop immediately as snowfields expand and thicken. In northern Quebec and Labrador, the population on interior lakes would have to migrate. Even the coastal regions of Hudson Bay and Hudson Strait would probably become uninhabitable due to constant snow accumulation throughout the year. In north central Quebec-Labrador, the Canadian Hydro power plants would be adversely affected by the presence of the perennial snowfields to the north, which would anchor low-pressure storm systems along the west side of Baffin Bay and bring Arctic air southward over the catchment areas throughout the summer. Under cloudier skies, this could make complete summer melting of snow uncertain, and in the worst case could leave perennial snowfields developing into ice sheets over the central catchment areas also. The economic loss to Canada and to the northeastern United States where much of the hydro power is marketed would be a serious matter. Still farther south in Quebec and the Maritimes, heavy rains and deep winter snows would shorten the growing season and make crop cultivation difficult. New England would be similarly affected, but to a lesser extent. Figure 63 suggests the likely areas that would be most severely affected by the increase in precipitation after the switch to deepwater formation within Baffin Bay occurs.

For northern Europe our speculation can be guided by the pollen records when the last ice age was initiated. The disappearance of warmth-loving species such as linden trees, ivy, and holly suggest a climate of colder winters and cooler summers. A dryer climate is also likely. Many crops now grown in northern Europe would fail. A climate somewhat more severe than during the Little Ice Age of the 1700's is likely, and considerable hardship in daily living would result. When the packice on the Arctic Ocean vanishes, the storms that would then follow the fringes of the ocean around the pole would very likely initiate local or regional ice-sheet growth in northern Russia and Siberia, Scandinavia, and possibly Alaska and the Inuit islands of

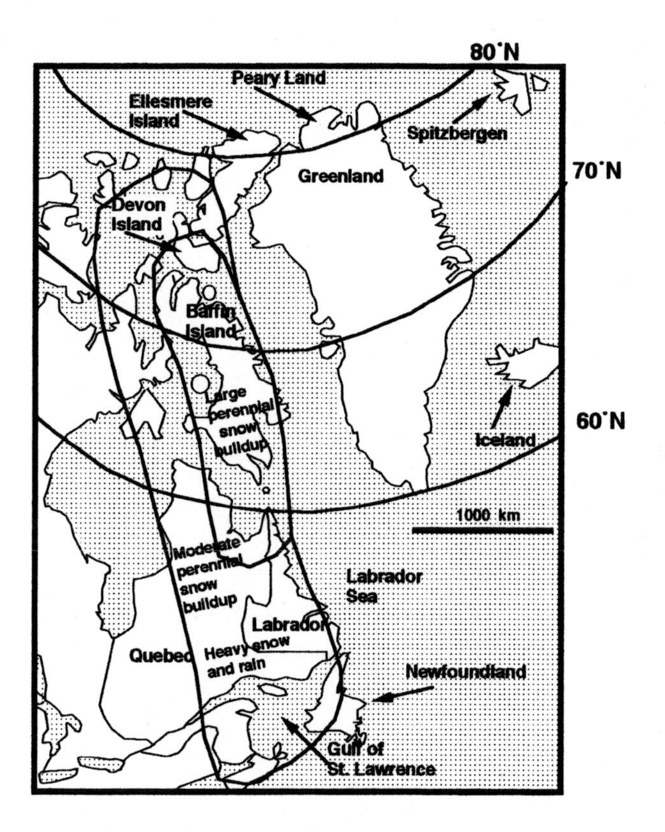

Figure 63: Precipitation increases after the ice age begins. Deepwater formation begins in Baffin Bay and the storm systems become locked onto the strip of contrasting temperatures between open water in Baffin Bay and the cold lands to the west.

northern Canada. Many of these regions have valuable natural resources. The recovery of these resources could become quite difficult, and there, too, the lands may become uninhabitable.

Against these dreary disadvantages, there are two distinct consequences that can be viewed positively:

(1) The added amount of deepwater formation in Baffin Bay would reduce the rate of increase of CO_2 in the atmosphere, at least over the next century or two. This would occur because CO_2 dissolves in the sea-surface water and is carried into the deep ocean when deepwater sinks. There is an equal amount of water upwelling elsewhere in the world, but it is "old" water that sank when there was less CO_2 in the atmosphere and therefore does not add as much CO_2 to the atmosphere as is taken into the deep ocean by water that sinks today. Although this would reduce the rate of global warming, the magnitude of the reduction is not known.

(2) The other consequence is the fall in sealevel that would result if the new glaciation in the cold interior of Canada withdraws water from the ocean faster than global warming adds it due to melting of Antarctic and Greenland ice in warmer coastal areas. On Barbados, the average rate of world sealevel fall in the first 4000 years of ice sheet growth that began 120,500 years ago was about 200 cm per thousand years. The first 400-500 years of rapid ice accumulation may have been even greater than that rate. If so, in the first 50 years, sealevel could fall by 10 cm or more, if no melting of ice or thermal expansion of the ocean occurred due to global warming. If current sealevel is rising at a rate of several millimeters every decade, the onset of new ice-sheet growth in Canada might arrest or reverse the trend of rising sealevel today. This would certainly benefit port cities like Venice, Italy, that have serious tidal flooding problems, and Pacific atolls and low-lying island countries like the Seychelles where even a minor rise will make the islands uninhabitable. This desirable result would occur on a 50 year timescale. On a 500 year timescale it is possible

that a continuing development of glaciation could provide a cooling counterbalance to the effects of carbon dioxide warming in the Northern Hemisphere, thus preventing the melting of the Greenland Ice Sheet, and flooding of world coastal areas as the CO_2 concentration increases toward 2000 parts per million. But the fate of the ice sheet that covers the Antarctic continent would be uncertain.

Would global greenhouse warming reverse the glacial growth process, once it becomes firmly established ? With the chaotic interactions of the climate factors, it is probably foolish to venture an opinion. It may be that General Circulation Model capability will advance enough that an answer can be given by modeling with a reasonable degree of confidence. But in the end, Nature will have her way. Twenty-first century humanity is building an extremely effective worldwide greenhouse, we are driving the Mediterranean salinity upward to levels not seen in the last half-million years, and Mother Nature has no statute of limitations with regard to environmental abuse.

This quest for understanding of ice-age climate change rests for the moment. The significance of the salty Mediterranean outflow now is clear, and it provides us with a better insight on how variations of incoming solar energy have changed Pleistocene climates, how the last ice age began and ended, and how it is that Earth sometimes has warm interglacial periods and other times not. We may even see a rather different near-future climate a little more clearly. But one thing we know for sure: The climate changes now in progress will bring great challenges to our children and their descendants for a long time to come.

* * *

Appendix A: Astronomical Basics

Climate history would have been simpler if Earth's polar axis were perpendicular to the plane of the orbit around the sun, and if the orbit were circular. However, today the polar axis is inclined about 23.4° to the plane of the orbit, and the orbit is elliptical with the sun offset from the center at one focus of the ellipse (Fig. 64). Because of the 23.4° inclination of the axis we have seasonal climates, and because of the elliptical orbit, the Northern and Southern Hemispheres take turns receiving greater and lesser amounts of incident solar energy in the summer season over a cyclic interval of about 23,000 years as the summertime distance from the sun changes. This energy input tends to make summers warmer at some times and cooler at others about 11,500 years later.

The changing position of summer on the earth's orbit is caused by the lack of spherical symmetry in the earth's shape, which is flattened at the poles due to the daily rotation. The difference between the gravitational tug of the sun on the nearest side and the farthest side of the earth's equatorial bulge is very small, but it is enough to continually shift the axis of the rotation in a direction perpendicular to the pull of the sun's gravity. The moon exerts a similar effect in the same direction, and the combined forces make the axis slowly wobble like the axis of a spinning top. This is known as a precession effect, and it has a cycle of about 26,000 years. But other gravitational effects slowly rotate the axis of the earth's elliptical orbit in the orbital plane, reducing the effective insolation precessional cycle to 23,000 years, more or less.

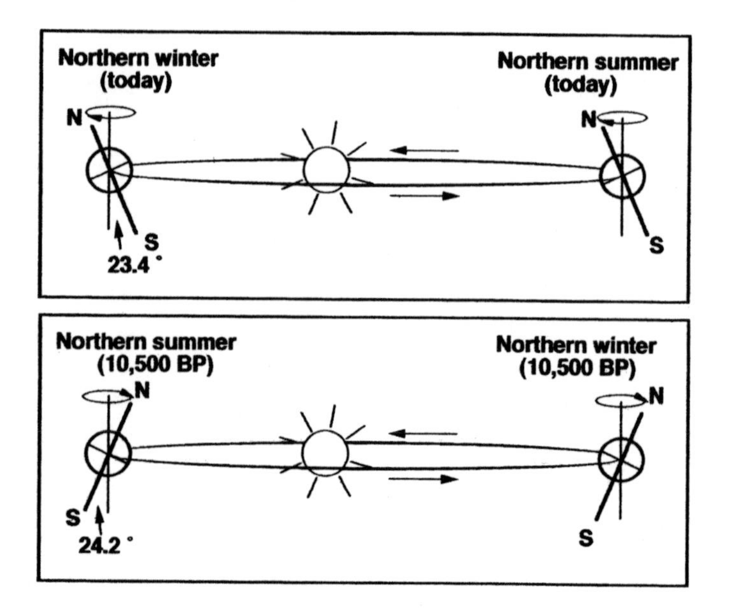

Figure 64: Effects of precession of the equinoxes. An almost edge-on view of Earth's orbit with the major axis of the ellipse in the plane of the paper. The rotational wobble of Earth's polar axis causes the summer solstice to occur at progressive points along the orbit around the sun. About 10,500 years ago Northern Hemisphere summer occurred near the point of Earth's closest approach to the sun. Since then the polar axis has swiveled around so that summer now occurs near the most distant point from the sun, which is the point in the orbit that is reached about July 6.

Because the precession effect on solar insolation is a function of varying distance from the sun, the increases or decreases of received solar energy occur at the same time at both Northern and Southern Hemisphere latitudes. The size of precessional insolation changes depends on the eccentricity of the orbit, which varies significantly on a scale of many tens of thousands of years. At present, the eccentricity is diminishing. Twenty thousand years from today the orbit will be nearly circular and the precession effect on insolation will be quite small.

The tilt, the inclination of the polar axis to the plane of the orbit, tends to make Northern Hemisphere summer insolation high when Southern insolation is low, and vice versa. The tilt has an amplitude of about 2.4° centered at a nominal 23.3° angle to the plane of the earth's orbit (Fig. 65), with a cycle of about 41,000 years. At high latitudes a difference of 2° in tilt can make a noticeable difference in

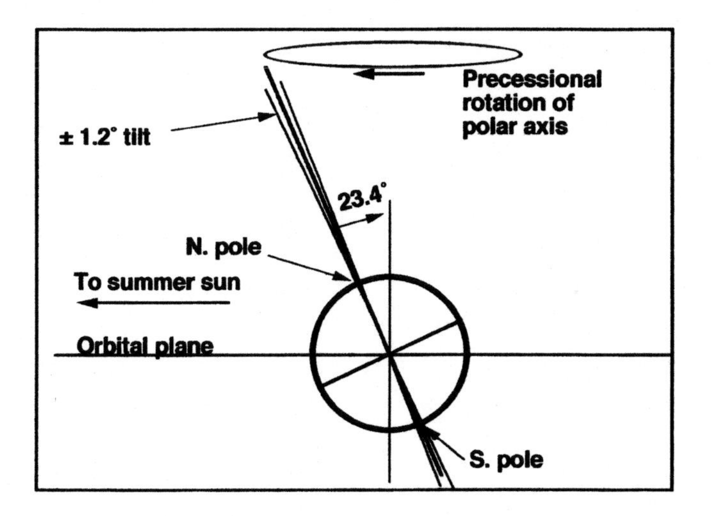

Figure 65: Tilt variation of Earth's polar axis. Today the angle of inclination is about 23.4°, but over a cycle of about 41,000 years the angle changes by as much as ±1.2° from its average position. The effect on insolation variations is strongest at the poles, and nearly vanishes at the equator.

the incident energy on each square meter of the earth's surface (or more precisely, the surface of the top of the atmosphere) because the sun is low in the sky even at noon at midsummer. Near the equator at low latitudes the tilt variation has very little effect.

The amplitudes of the resulting insolation changes are, of course, dependent on latitude, and are highly variable over tens and hundreds of thousands of years because the eccentricity of the orbit and tilt of the axis change independently. In quite rare cases, the high-latitude midsummer-day insolation extremes could be different by over 20%. Midsummer extremes are less at low latitudes because the tilt effect is small there. However, the differences between extremes of the caloric summer half-year averages at each latitude that are customarily used for speculations on the Milankovitch hypothesis are nearly always less than about 8%. Note that the caloric summer is defined as that half of the year when every day has more insolation received at the top of the atmosphere than any other day in the winter half.

The magnitudes of these insolation variations are not large, but it is inevitable that they would have some effect on the climate. An important earth-science question of our times is: How much of the observed climate change in the complex real world is due to the orbital factors, and how do the insolation variations affect the circulation of the atmosphere and ocean ?

Appendix B: The bi-polar seesaw

In considering the possibility of future climate change, it would be unwise to ignore the evidence for successive warmings and coolings in the North Atlantic over the last 10,000 years. These cycles have a length of roughly 1500 years. Among the papers that have been published on the subject is an extensive report on the cycles in foraminiferal data from core V28-14 between Iceland and Greenland, and core V29-191 west of Ireland (G. Bond et al., 1997). The consistent temperature changes at these locations have an average cyclic duration of 1470±530 years. The cause of the oscillation is thought to be alternating and opposing increases and decreases of deepwater formation in the high latitudes of the North Atlantic and the Southern Ocean, and has been modeled in detail by Seidov and colleagues (2001).

In simplified form, one can imagine how such an oscillation might work based on ideas discussed in earlier chapters. If an increase in North Atlantic Deep Water (NADW) formation supplied added salt to the water around Antarctica, the additional salinity might increase the rate of Antarctic Bottom Water (AABW) formation. AABW underlies the North Atlantic water and may be diffusing up into the abyssal currents that move northward off the coast of Europe in accord with the coriolis effect. If this upward diffusion occurs, as suggested by the vertical trends of salinity and silica concentration, the salinity of the Mediterranean outflow mixture would be influenced by the rate of relatively low-salinity AABW production. When AABW production increases, the salinity of the mixture reaching the surface in the northern gyre would become less, thus tending to eliminate the original increase in NADW formation. The elimination effect would be delayed by several hundred years because of transit times and mixing times in the Northern and Southern Oceans, and the delay is, of course, the cause of the oscillation. This mechanism would probably be an over-simplification, even if it is valid, because the length of the cycle is quite variable.

Nevertheless, the cyclic changes exist, and it is of interest to try to see where we stand in the cycle today. The last maximum of cyclic warmth can be assigned to the period of Norse settlement on Greenland about 1000 AD. If the half-cycle is 750 years, the cyclic minimum would have been about 1750 AD during the extended Little Ice Age. For example, an extreme winter occurred in 1789 when people walked on the ice from Staten Island to Manhattan (Imbrie and Imbrie, 1979), which seems almost incredible today. With these two points on the timescale, the next maximum would occur around 2500 AD. We are now 250 years into the warming.

It may be comforting that the last interval of maximum warmth was caused by an increase in conveyor-belt strength, perhaps even with a little deepwater forming in northern Baffin Bay, thus enabling the Norse to sail around Greenland. The bi-polar seesaw and the increasing saline outflow from the Mediterranean should both tend to counteract the shutdown of the conveyor-belt predicted by the Manabe and Stouffer model, and may therefore prolong the present equable European climate. It seems likely, however, that the trends of the bi-polar seesaw and the Mediterranean outflow would combine to ensure the onset of major glaciation in Canada in the near future.

Bibliography

Adam, D.P. 1976. Hudson Bay, Lake Zissaga and the growth of the Laurentide Ice Sheet. *Nature* 261:72-73.

Adkins, J.F., E.A. Boyle, L. Keigwin, and E. Cortijo. 1997. Variability of North Atlantic thermohaline circulation during the last glacial period. *Nature* 390:154-156.

Aksu, A.E. 1985. Climatic and oceanographic changes over the past 400,000 years: evidence from deep-sea cores on Baffin Bay and Davis Strait. In: *Quaternary environments, eastern Canadian Arctic, Baffin Bay and western Greenland*. J.T. Andrews (ed.), 181-209. Boston: Allen and Unwin.

American Institute of Physics Handbook. 1957. D.E. Gray et al. (eds.) New York: McGraw-Hill.

Andrews, J.T. and M.A.W. Mahaffy. 1976. Growth rate of the Laurentide Ice Sheet and sea level lowering (with emphasis on the 115,000 BP sea level low). *Quaternary Research* 6:167-183.

Andrews, J.T. 1998. Abrupt changes (Heinrich events) in late Quaternary North Atlantic marine environments: a history and review of data and concepts. *Journal of Quaternary Science* 13(1):3-16.

Arkhipov, S.A., J. Ehlers, R.G. Johnson and H.E. Wright, Jr. 1995. Glacial drainage towards the Mediterranean during middle and Late Pleistocene. *Boreas* 24:196-206.

Armand, L.R. 2000. An ocean of ice - Advances in the estimation of past sea ice in the Southern Ocean. *GSA Today* 10(No. 3):1-7.

Bard, E., F. Rostek, J.-L. Turon, and S. Gendreau. 2000. Hydrological impact of Heinrich events in the subtropical northeast Atlantic. *Science* 289:1321-1324.

Barry, R.G., J.T. Andrews, and M.A. Mahaffy. 1975. Continental ice sheets: Conditions for growth. *Science* 190:979-981.

Bender, M.L., R.G. Fairbanks, F.W. Taylor, R.K. Matthews, J.G. Goddard, and W.S. Broecker. 1979. Uranium-series dating of the Pleistocene reef tracts of Barbados, West Indies. *GSA Bulletin* 90:577-594.

Berger, A.L. 1978. Long term variations of caloric summer insolation resulting from the Earth's orbital elements. *Quaternary Research* 9:139-167 (and tabulated results).

Blake Jr., W. 1966. End moraines and deglaciation chronology in northern Canada, with special reference to southern Baffin Island. *Geological Survey of Canada, Paper 66-26*.

Blanc, P.-L. and J.-C. Duplessy. 1982. The deep-water circulation during the Neogene and the input of the Messinian salinity crisis. *Deep-Sea Research* 29:1391-1414.

Bloom, A.L., W.S. Broecker, J.M.A. Chappell, R.K. Matthews, and K.J. Mesolella. 1974. Quaternary sealevel fluctuations on a tectonic coast: New 230Th/234U dates from the Huon Peninsula, New Guinea. *Quaternary Research* 4:185-205.

Bodén, P., R.G. Fairbanks, J.D. Wright, and L.H. Burckle. 1997. High resolution stable isotope records from southwest Sweden: The drainage of the Baltic Ice Lake and Younger Dryas ice-margin oscillations. *Paleoceanography* 12:39-49.

Bond, G., W. Broecker, S. Johnsen, J. McManus, L. Labeyrie, J. Jouzel, and G. Bonani. 1993. Correlations between climate records from North Atlantic sediments and Greenland ice. *Nature* 365:143-147.

Bond, G., and R. Lotti. 1995. Iceberg discharges into the North Atlantic on millennial time scales during the last glaciation. Science 267:1005-1010.

Bond, G., W. Showers, M. Cheseby, R. Lotti, P. Almasi, P. deMenocal, P. Priore, H. Cullen, I. Hajdas, and G. Bonani. 1997. A pervasive millennial-scale cycle in North Atlantic Holocene and glacial climates. *Science* 278:1257-1266.

Broecker, W.S. 1999. What if the conveyor belt were to shut down ? Reflections on a possible outcome of the great global experiment. *GSA Today* 9(No.1).

Bryden, H.L. and T.H. Kinder. 1991. Steady two-layer exchange through the Strait of Gibraltar. *Deep-Sea Research* 38:5445-5463.

Chappell, J. 1974. Geology of coral terraces, Huon Peninsula, New Guinea: a study of Quaternary tectonic movements and sea-level changes. *Geological Society of America Bulletin* 85:553-570.

Chappell, J. and N.J. Shackleton. 1986. Oxygen isotopes and sea level. *Nature* 324:137-140.

Clark, P.U., R.B. Alley, L.D. Keigwin, J.M. Licciardi, S.J. Johnsen, and H. Wang. 1996. Origin of the first global meltwater pulse following the last glacial maximum. *Paleoceanography* 11:563-577.

Clark, P.U., S.J., Marshall, G.K.C. Clarke, S.W. Hostetler, J.M. Licciardi, and J.T. Teller. 2001. Freshwater forcing of abrupt climate change during the last glaciation. *Science* 293:283-287.

CLIMAP Project Members. 1976. The surface of the ice age Earth. *Science* 191:1131-1137.

Croll, J. 1875. *Climate and time*. New York:Appleton.

Dansgaard. W., and J. Tauber. 1969. Glacier oxygen-18 content and Pleistocene ocean temperatures. *Science* 166:494-502.

Deacon, G., The Antarctic Ocean. 1977. *Interdisciplinary Science Reviews* 2:109-123.

Denton, G.H., and T.J. Hughes. (eds.). 1981. *The Last Great Ice Sheets*. New York: Wiley.

Dickson, R.R., J. Meincke, S.-A. Malmberg, and A.J. Lee. 1988. The "Great Salinity Anomaly" in the northern North Atlantic 1968-1982. *Progress in Oceanography* 20:103-151.

Duplessy, J.D., L. Labeyrie, M. Arnold, M. Paterne, J. Duprat, and T.C.E. Weering. 1992. Changes in surface salinity of the North Atlantic Ocean during the last deglaciation. *Nature* 358:485-487.

Dyke, A.S., and V.K. Prest. 1986. Late Wisconsinan and Holocene retreat of the Laurentide Ice Sheet. *Geological Survey of Canada*, Map 1702A.

Edwards, R.L., J.H. Chen, and G.J. Wasserburg. 1986/97. 238U-234U-230Th-232Th systematics and precise measurement of time over the past 500,000 years. *Earth and Planetary Science Letters* 81:175-192.

Ehlers, J., K.-D. Meyer, and H.-J. Stephan. 1984. The pre-Weichselian glaciations of north-west Europe. *Quaternary Science Reviews* 3:1-40.

Emiliani, C., Pleistocene temperatures. 1955. *Journal of Geology* 63:538-578.

Esat, T.M., M.T. McCulloch, J. Chappell, B. Pillans, and A. Omura. 1999. Rapid fluctuations in sea level recorded at Huon Peninsula during the penultimate deglaciation. *Science* 283:197-201.

Fairbanks, R.G. 1990. The age and origin of the "Younger Dryas climate event" in Greenland ice cores. *Paleoceanography* 5:937-948.

Field, M.H., B. Huntley, and H. Müller. Eemian climate fluctuations observed in a European pollen record. 1994. *Nature* 371:779-783.

Fillon, R.H. 1985. Northwest Labrador Sea stratigraphy, sand input and paleoceanography during the last 160,000 years. 1985. In: *Quaternary Environments, Eastern Canadian Arctic, Baffin Bay and Western Greenland.*. J.T. Andrews (ed.), 212-247. Boston: Allen and Unwin.

Gallup, C.D., R.L. Edwards, and R.G. Johnson. 1994. The timing of high sea levels over the past 200,000 years. *Science* 263:796-800.

Gasser, M.M., and F. El-Gamal, Aswan High Dam: Lessons learnt and on-going research. 1994. *Water power and Dam Construction*, January:35-39.

Greatbatch, R.J., and J. Xu. 1993. On the transport of volume and heat through sections across the North Atlantic: Climatology and the pentads 1955-1959, 1970-1974. *Journal of Geophysical Research* 98:10125-10143.

Gunnerson, C.G., and E. Özturgut. 1974. The Bosporus. In: *The Black Sea - Geology, Chemistry and Biology, Memoir 20.* Degens and Ross (eds.), 99-114. American Association of Petroleum Geologists.

Hays, J.D., J. Imbrie, and N.J. Shackleton. 1976. Variations in the Earth's orbit: Pacemaker of the Ice Ages. *Science* 194:1121-1132.

Hays, J.D., J.A. Lozano, N. Shackleton, and G. Irving. 1976b. Reconstruction of the Atlantic and western Indian Ocean sectors of the 18,000 B.P. Antarctic Ocean. In: *Investigation of Late Quaternary Paleoceanography and Paleoclimatology.* R.M.Cline and J.D. Hays (eds.), 327-332. *Geological Society of America Memoir 145.*

Hearty, P.J. 1987. New data on the Pleistocene of Mallorca. *Quaternary Science Reviews* 6:245-257.

Hearty, P.S., P. Kindler, H. Chang, and R.L. Edwards. 1999. A +20 m middle Pleistocene sea-level highstand (Bermuda and the Bahamas) due to partial collapse of Antarctic ice. *Geology* 27:375-378.

Hebbeln, D., T. Dokken, E.S. Andersen, M. Hald, and A. Elverhoi,. 1994. Moisture supply for northern ice-sheet growth during the Last Glacial Maximum. *Nature* 370:357-360.

Heinrich, H. 1988. Origin and consequences of cyclic rafting in the Northeast Atlantic Ocean during the past 130,000 years. *Quaternary Research* 29:143-152.

Helgason, J, and R.A. Duncan. 2001. Glacial-interglacial history of the Skaftafell region, southeast Iceland, 0-5 Ma. *Geology* 29:179-182.

Herweijer, J.P., and J.W. Focke. 1978. Late Pleistocene depositional and denudation history of Aruba, Bonaire, and Curaçao (Netherlands Antilles). *Geologie en Mijnbow* 57:177-187.

Holand, H.R. 1940. *Westward from Vinland.* New York:Duell, Sloan & Pearce.

Hollin, J.T., and P.J. Hearty. 1990. South Carolina interglacial sites and stage 5 sealevels. *Quaternary Research* 33:1-17.

Hoyle, F. 1981. *Ice: A chilling scientific forecast of a new ice age.* Kent, England:New English Library.

Hsü, K.J.. and nine others. 1977. History of the Mediterranean salinity crisis. *Nature* 267:399-403.

Hurrell, J.W., Decadal trends in the North Atlantic Oscillation: Regional temperatures and precipitation. 1995. *Science* 269:676-679.

Imbrie, J. and K.P. Imbrie. 1979. *Ice Ages: solving the mystery*. Cambridge, Mass.:Harvard U. Press.

Ingstad, H. 1969. *Westward to Vinland*. New York:St. Martins Press.

Ingstad, H., and A.S. Ingstad. 2001. *The Viking Discovery of America*. St. Johns, NF:Breakwater Books Ltd.

Johnsen, S.J., W. Dansgaard, and J.W.C. White. 1989. The origin of Arctic precipitation under present and glacial conditions. *Tellus 41B*: 452-468.

Johnsen, S.J., H.B. Clausen, W. Dansgaard, N.S. Gundestrup, C.U. Hammer, and H. Tauber. 1995. The Eem stable isotope record along the GRIP ice core and its interpretation. *Quaternary Research* 43:117-124.

Johnsen, S.J., H.B. Clausen, W. Dansgaard, and twelve others. 1997. The $\delta^{18}O$ record along the Greenland Ice Core Project deep ice core and the problem of possible Eemian climatic instability. *Journal of Geophysical Research* 102(No.C12):26,397-26,410.

Johnson, R.G., and B.T. McClure. 1976. A model for Northern Hemisphere continental ice sheet variation. *Quaternary Research* 6:325-353.

Johnson, R.G. 1982. Brunhes-Matuyama magnetic reversal dated at 790,000 yr B.P. by marine-astronomical correlations. *Quaternary Research* 17:135-147.

Johnson, R.G., and S.-E. Lauritzen. 1995. Hudson Bay-Hudson Strait jokulhlaups and Heinrich events: a hypothesis. *Palaeogeography, Palaeoclimatology, Palaeoecology* 117:123-137.

Johnson, R.G., Ice age initiation by an ocean-atmospheric circulation change in the Labrador Sea. 1997. *Earth and Planetary Science Letters* 148:367-379.

Johnson, R.G. 2001. Last interglacial seastands on Barbados and an early anomalous deglaciation timed by differential uplift. *Journal of Geophysical Research* 106(No. C6):11543-11551.

Jones, G.A., and L.D. Keigwin. 1988. Evidence from the Fram Strait (78°N) for early deglaciation. *Nature* 336:56-59.

Kallel, N., J.-C. Duplessy, L. Labeyrie, M. Fontugne, M. Paterne, and M. Montacer, Mediterranean pluvial periods and sapropel formation over the last 200,000 years. 2000. *Palaeogeography Palaeoclimatology Palaeoecology* 157:45-58.

Karner, D.G., and R.A. Muller. 2000. A causality problem for Milankovitch. *Science* 288:2143-2144.

Kaufman, D.S., G.H. Miller, J.A. Stravers, and J.T. Andrews. 1993. Abrupt early Holocene (9.9-9.6ka) ice-stream advance at the mouth of Hudson Strait, Arctic Canada. *Geology* 21:1063-1066.

Keigwin, L.D., Jr., Pliocene closing of the Isthmus of Panama, based on biostratigraphic evidence from nearby Pacific Ocean and Caribbean Sea cores. 1978. *Geology* 6:630-634.

Klein, W.H., Principal tracks and mean frequencies of cyclones and anticylones in the Northern Hemisphere. 1957. *Research Paper No. 40*. Washington, D.C.:U. S. Weather Bureau.

Koerner, R.M., Ice core evidence for extensive melting of the Greenland ice sheet in the last interglacial. 1989. *Science* 244:964-968.

Koerner, R.M., J.C. Bourgeois, and D.A. Fisher,. 1988. Pollen analysis and discussion of time-scales in Canadian ice cores. *Annals of Glaciology* 110:85-91.

Krabill, W. and nine others. 2000. Greenland ice sheet: High elevation balance and peripheral thinning. *Science* 289:428-430.

Lauritzen, S.-E.. 1995. High-resolution paleotemperature proxy record for the last interglaciation based on Norwegian speleothems. *Quaternary Research* 43:133-146.

Lehman, S.J., and L.D. Keigwin. 1992. Deep circulation revisited, *Nature* 358:197-198.

Libby, W.F. 1952. *Radiocarbon Dating*. Chicago:University of Chicago Press.

MacAyeal, D.R. 1993. Binge/purge oscillations of the Laurentide ice sheet as a cause of the North Atlantic's Heinrich events. *Paleoceanography* 8:775-784.

Manabe, S., and R.J. Stouffer. 1993. Century-scale effects of increased atmospheric CO_2 on the ocean-atmosphere system, *Nature* 364:215-218.

Martinson, D.G., N.G. Pisias, J.D. Hays, J.I. Imbrie. T.C. Moore Jr., and N.J. Shackleton. 1987. Age dating and the orbital theory of the ice ages: development of a high-resolution 0 to 300,000-year chronostratigraphy. *Quaternary Research* 27:1-29.

McNeill, D.F., A.G. Coates, A.F. Budd, and P.F. Borne. 2000. Integrated paleontologic and paleomagnetic stratigraphy of the upper Neogene deposits around Limon, Costa Rica: A coastal emergence record of the central American Isthmus. *GSA Bulletin* 112:963-981.

Milankovitch, M. 1930. *Mathematische klimalehre und astronomische theorie der klimaschwankungen*. In: *Handbuch der Klimatologie*, I (A). W. Koppen and R. Geiger (eds.),1-176 Berlin:Gebruder Borntraeger.

Miller, G.H., and A. deVernal. 1992. Will greenhouse warming lead to Northern Hemisphere ice sheet growth ? *Nature* 355:244-246.

Muerdter, D.R. 1984. Low-salinity surface water incursions across the Strait of Sicily during Late Quaternary sapropel intervals. *Marine Geology* 58:401-414.

Neumann, A.C. and P.J. Hearty. 1996. Rapid sea-level changes at the close of the last interglacial (substage 5e) recorded in Bahamian island geology. *Geology* 24:775-778.

Opdyke, N.D. 1972. Paleomagnetism of deep-sea cores. *Review Geophysics and Space Physics* 10:213-249.

Otto, L., and H.M. van Aken. 1996. Surface circulation in the northeast Atlantic as observed with drifters. *Deep Sea Research I,* 43:467-499.

Pisias, N.G., D.G. Martinson, T.C. Moore, Jr., N.J. Shackleton, W. Prell, J. Hays. and G. Bodén. 1984. High resolution stratigraphic correlation of benthic oxygen isotopic records spanning the last 300,000 years. *Marine Geology 56*:119-136.

Pollard, R.T., and S. Pu. 1985. Structure and circulation of the upper Atlantic Ocean northeast of the Azores. *Progress in Oceanography* 14:443-462.

Reid, J.L. 1978. On the mid-depth circulation and salinity field in the North Atlantic Ocean. *Journal of Geophysical Research* 83:5063-5067.

Reid, J.L. 1979. On the contribution of the Mediterranean Sea outflow to the Norwegian-Greenland Sea. *Deep-Sea Research* 26:1199-1223.

Reid, J.L. 1994. On the total geostrophic circulation of the North Atlantic Ocean: Flow patterns, tracers and transports. *Progress in Oceanography* 33:1-92.

Rind, D., D. Peteet, and G. Kukla. 1989. Can Milankovitch orbital insolation initiate growth of ice sheets in a general circulation model ? *Journal of Geophysical Research* 94:12851-12871.

Rohling, E.J., and H.L. Bryden. 1992. Man-induced salinity and temperature increases in Western Mediterranean deep water. *Journal of Geophysical Research* 97(No. C7):11191-11198.

Rossignol-Strick, M.. 1983. African monsoons, an immediate climate response to orbital insolation, *Nature* 304:46-49.

Rossignol-Strick, M. 1985. Mediterranean sapropels, an immediate response of the African monsoon to variation of insolation. *Palaeogeography, Palaeoclimatology, Palaeoecology* 49:237-263.

Rossignol-Strick, M., M. Paterne, F.C. Bassinot, K.-C. Emeis, & G.J. De Lange. 1998. An unusual mid-Pleistocene monsoon period over Africa and Asia. *Nature* 392:269-272.

Rothrock, D.A., Y. Yu, and G.A. Maykut. 1999. Thinning of Arctic sea-ice cover. *Geophysical Research Letters* 26:3469-3472.

Ruddiman, W.F. 1977. Late Quaternary deposition of ice-rafted sand in the subpolar North Atlantic (lat 40°N to 65°N). *Geological Society of America Bulletin* 88:1813-1827.

Ruddiman, W.F., and A. McIntyre. 1979. Warmth of subpolar North Atlantic Ocean during Northern Hemisphere ice-sheet growth. *Science* 204:173-175.

Ruddiman, W.F., A. McIntyre, J. Niebler-Hunt, and J.T. Durazzi. 1980. Oceanic evidence for the mechanism of rapid Northern Hemisphere glaciation. *Quaternary Research* 13:33-64.

Ruddiman, W.F., and A. McIntyre. 1981. Oceanic mechanisms for amplification of the 23,000 year ice-volume cycle. *Science* 212:617-627.

Ruddiman, W.F., and A. McIntyre. 1982. Severity and speed of Northern Hemisphere glaciation pulses: The limiting case ? *Geological Society of America Bulletin 93:*1273-1279.

Ryan, W., and Pitman, W. 1998. *Noah's Flood.* New York: Simon and Schuster.

Ryan, W.F.B., W.C. Pitman III, C.O. Major, K. Shimkus, V. Moskalenka, G.A. Jones, P. Dimitrov, N. Gorür, M. Sakinç, and H. Yüce. 1997. An abrupt drowning of the Black Sea shelf. *Marine Geology* 138:119-126.

Schönfeld, J., and R. Zahn. 2000. Late Glacial to Holocene history of the Mediterranean outflow. Evidence from benthic foraminiferal assemblages and stable isotopes at the Portuguese margin, *Palaeogeography, Palaeoclimatology, Palaeoecology* 159:85-111.

Seidenkrantz, M.-S. 1993. Benthic and foraminiferal evidence for "Younger Dryas-style cold spell" at the Saalian-Eemian transition, Denmark. *Palaeogeography, Palaeoclimatology, Palaeoecology* 102:103-120.

Seidov, D., B.J. Haupt, E.J. Barron, and M. Maslin. 2001. Ocean bi-polar seesaw and climate: Southern versus Northern meltwater impacts. In: *The oceans and rapid climate change: Past, present and future*. D. Seidov, B.J. Haupt, and M. Maslin (eds.) *Geophysical Monograph 126.* Washington D.C.: American Geophysical Union.

Shackleton, N.J., and N.D. Opdyke. 1977. Oxygen isotope and palaeomagnetic evidence for early Northern Hemisphere glaciation, *Nature* 270:216-219.

Shackleton, N.J., A. Berger, and W.R. Peltier. 1990. An alternative astronomical calibration of the lower Pleistocene timescale based on ODP site 677. *Transactions of the Royal Society of Edinburgh: Earth Sciences* 81:251-261.

Skelton, R.A., T.E. Marston, and G.D. Painter. 1965. *The Vinland Map and theTartar Relation*. New Haven: Yale University Press.

Stea, R.R., R. Boyd, O. Costello, G.B.J. Fader, and D.B. Scott. 1996. In: *Late Quaternary Palaeoceanography of the North Atlantic Margins*. J.T. Andrews, et al., (eds.), 77-101. *Geological Society Special Publication No. 111*.

Stravers, J.A., G.H. Miller, and D.S. Kaufman. 1992. Late glacial ice margins and deglacial chronology for southeastern Baffin Island and Hudson Strait, eastern Canadian Arctic, *Canadian Journal of Earth Science 29:*1000-1017.

Street-Perrott, F.A., and R.A. Perrott. 1993. Holocene vegetation, lake levels, and climate of Africa. In: *Global Climates Since the Last Glacial Maximum,* H.E. Wright Jr. et al. (eds.), 318-356. Minneapolis: University of Minnesota Press.

Stuiver, M.. 1980. Solar variability and climatic change during the current millennium. *Nature* 286:868-871.

Stuiver, M., P.J. Reimer, E. Bard, J.W. Beck, G.S. Burr, K.A. Hughen, B. Kromer, G. McCormac, J. van der Plicht, and M. Spurk. 1998. INTCAL98 radiocarbon age calibration, 24,000-0 cal BP. *Radiocarbon* 40:1041-1083.

Thunell, R.C., and D.F. Williams. 1982. Paleoceanographic events associated with termination II in the Eastern Mediterranean. *Oceanologica Acta* 5:229-233.

Upham, W. 1880. Preliminary report on the geology of central and western Minnesota. *Minnesota Geological and Natural History Survey, 8th annual report, yr 1879*, 70-125.

Vernekar, A.D. 1972. Long-period global variations on incoming solar radiation. *Meteorological Monographs 12, No. 34,* American Meteorological Society.

Walker, J.C.G., and J.R. Kasting. 1992. Effects of fuel and forest conservation on future levels of atmospheric carbon dioxide. *Palaeogeography, Palaeoclimatology, Palaeoecology* 97 (global and planetary change section):151-189.

Wegener, A. 1966. Origin of continents and oceans (English translation). New York: Dover Publications.

Weyl, P.K.. 1968. The role of oceans in climate change: A theory of the ice ages. In: *Causes of Climate Change*. J.M. Mitchell (ed.), 37-62. *Meteorological Monographs* 8(No.30). American Meteorological Society.

Winograd, I.J., J.M. Landwehr, K.R. Ludwig, T.B. Coplen, and A.C. Riggs. 1997. Duration and structure of the past four interglaciations. *Quaternary Research* 48:141-154.

Worthington, L.V. 1976. *On the North Atlantic Circulation*. Baltimore: Johns Hopkins University Press.

Yan, Z., and N. Petit-Maire. 1994. The last 140 ka in the Afro-Asian arid/semi-arid transitional zone. *Palaeogeography, Palaeoclimatology, Palaeoecology* 110:217-233.

Index

* * * *